AF553786

TEXTBOOK OF MICROPALAEONTOLOGY

TEXTBOOK OF MICROPALAEONTOLOGY

By
Dr. P.R. Yadav
Lecturer
Dept. of Zoology
D.A.V. College
Muzaffarnagar (U.P.)
(India)

Published by:

DISCOVERY PUBLISHING HOUSE
4383/4B, Ansari Road, Darya Ganj
New Delhi-110 002 (India)
Phone : +91-11-23279245; 23253475; 43596065
E-mail : discoverybooksindia@gmail.com
orderdphbooks@gmail.com
namitwasan9@gmail.com
web : www.discoverypublishinggroup.com

***First Edition:* 2010**
***Reprinted:* 2025**

ISBN: 978-81-8356-576-9

Textbook of Micropalaeontology

Printed at:
Infinity Imaging Systems
Delhi

Preface

Life on earth has suffered occasional mass extinctions at least since 542 million years ago. Although they are disasters at the time, mass extinctions have sometimes accelerated the evolution of life on earth. When dominance of particular ecological niches passes from one group of organisms to another, it is rarely because the new dominant group is "superior" to the old and usually because an extinction event eliminates the old dominant group and makes way for the new one.

The fossil record appears to show that the rate of extinction is slowing down, with both the gaps between mass extinctions becoming longer and the average and background rates of extinction decreasing. However, it is not certain if the actual rate of extinction has altered, since both of these observations could be explained in several ways.

The oceans may have become more hospitable to life over the last 500 million years and less vulnerable to mass extinctions dissolved oxygen became more widespread and penetrated to greater depths; the development of life on land reduced the run-off of nutrients and hence the risk of eutrophication and anoxic events; and marine ecosystems became more diversified so that food chains were less likely to be disrupted.

Reasonably complete fossils are very rare, most extinct organisms are represented only by partial fossils, and complete fossils are rarest in the oldest rocks. So paleontologists have mistakenly assigned parts of the same organism to different genera which were often defined solely to accommodate these finds – the story of Anomalocaris is an example of this. The risk of this mistake is higher for older fossils because these are often unlike parts of any living organism. Many of the "superfluous" genera are represented by fragments which are not found again and the "superfluous" genera appear to become extinct very quickly.

Biodiversity in the fossil record, which is "the number of distinct genera alive at any given time; that is, those whose first occurrence predates and whose last occurrence postdates that time"

A fairly swift rise from 542 to 400 million years ago; a slight decline from 400 to 200 million years ago, in which the devastating Permian–Triassic extinction event is an important factor; and a swift rise from 200 million years ago to the present.

– Author

Contents

CHAPTER–1

Introduction

Micropaleontology (also sometimes spelled as micro-palaeontology) is that branch of paleontology which studies microfossils. Microfossils are fossils generally not larger than four millimetres, and commonly smaller than one millimetre, the study of which requires the use of light or electron microscopy. Fossils which can be studied with the naked eye or low-powered magnification, such as a hand lens, are referred to as macrofossils. Obviously, it can be hard to decide whether or not some organisms should be considered microfossils, and so there is no fixed size boundary.

For example, some colonial organisms, such as bryozoa (especially the Cheilostomata) have relatively large colonies, but are classified on the basis of fine skeletal details of the tiny individuals of the colony. Most bryozoan specialists tend to consider themselves paleontologists, rather than micro-paleontologists, but many micropaleontologists also study bryozoa.

In another example, many fossil genera of Foraminifera, which are protists, are known from shells (called "tests") that were as big as coins, such as the genus *Nummulites*.

Microfossils are a common feature of the geological record, from the Precambrian to the Holocene. They are most common in deposits of marine environments, but also occur in brackish water, fresh water and terrestrial sedimentary deposits. While every kingdom of life is represented in the microfossil record, the most abundant forms are protist skeletons or cysts from the Chrysophyta, Pyrrhophyta, Sarcodina, acritarchs and chitinozoans, together with pollen and spores from the vascular plants.

Micropaleontology can be roughly divided into four areas of study on the basis of microfossil composition:

(a) calcareous, as in coccoliths and foraminifera;

(b) phosphatic, as in the study of some vertebrates;

(c) siliceous, as in diatoms and radiolaria; or

(d) organic, as in the pollen and spores studied in palynology.

This division reflects differences in the mineralogical and chemical composition of microfossil remains (and therefore in the methods of fossil recovery) rather than any strict taxonomic or ecological distinctions. Most researchers in this field, known as micropaleontologists, are typically specialists in one or more taxonomic groups.

Calcareous Microfossils

Calcareous [$CaCO_3$] microfossils include Coccoliths, Foraminifera, Calcareous dinoflagellates, and Ostracods (seed shrimp).

Phosphatic Microfossils

Phosphatic microfossils include Conodonts (tiny oral structures of an extinct chordate group), some scolecodonts ("worm" jaws), Shark spines and teeth, and other Fish remains (collectively called "ichthyoliths").

Siliceous Microfossils

Siliceous microfossils include Diatoms, Radiolaria, Silicoflagellates, phytoliths, some scolecodonts ("worm" jaws), and spicules.

Organic Microfossils

The study of organic microfossils is called palynology. Organic microfossils include pollen, spores, Chitinozoans (thought to be the egg cases of marine invertebrates), Scolecodonts ("worm" jaws), Acritarchs, Dinoflagellate cysts, and fungal remains.

Methods

Sediment or rock samples are collected from either cores or outcrops, and the microfossils they contain extracted by a variety of physical and chemical laboratory techniques, including sieving, density separation by centrifuge, and chemical digestion of the unwanted fraction. The resulting concentrated sample of microfossils is then mounted on a slide for analysis, usually by light microscope. Taxa are then identified and counted. The very large numbers of microfossils that a small sediment sample can often yield allows the collection of statistically robust datasets which can be subjected to multivariate analysis. A typical microfossil study will involve identification of a few hundred specimens from each of ten to a hundred samples.

Applications of Micropaleontology

Microfossils are especially noteworthy for their importance in biostratigraphy. Since microfossils are often extremely abundant, widespread, and quick to appear and disappear from the stratigraphic record, they constitute ideal index fossils from a biostratigraphic perspective. In addition, the planktonic and nektonic habits of some microfossils gives them the added bonus of appearing across a wide range of facies or paleoenvironments, and having near-global distribution making biostratigraphic correlation even more powerful and effective.

Microfossils also provide some of the most important records of global environmental change on long-timescales, particularly from deep-sea sediments. Across vast areas of the ocean floor the shells of planktonic micro-ogranisms sinking from surface waters provide the dominant source of sediment and they continuously accumulate (typically at rates of 20-50m/million years). Study of changes in assemblages of microfossils and of changes in their shell chemistry (e.g oxygen isotope composition) are fundamental to research on climate change in the geological past.

In addition to providing an excellent tool for sedimentary rock-body dating and for paleoenvironmental reconstruction – heavily used in both petroleum geology and paleoceanography – micropaleontology has also found a number of less orthodox applications, such as its growing role in forensic police investigation or in provenancing archaeological artefacts.

One field of paleontology that is often overlooked is micropaleontology, or the study of extremely small fossils. Micropaleontology, like microbiology, is mostly focused on extremely small organisms, such as one-celled organisms. Micropaleontology, however, can also encompass the very small larval stages that some larger animals go through as they grow.

Because micropaleontology examines such small organisms, which can only be properly examined through a microscope, it is often seen as a boring area of paleontology, much less exciting than digging up dinosaur bones. But micropaleontology is a very important field and is helping us to learn about the origins of life on Earth.

The oldest-known fossils to be found on Earth are sedimentary structures called stromatolites, found in Western Australia. They were made by cyanobacteria (also called "blue-green algae") that lived about 3.5 billion years ago, more than a billion years after the formation of the Earth. The geologic time period in which life was first starting out on Earth is known as the Precambrian Era (about 4 billion to 570 million years ago). The Cambrian Era (570 to 500 million years ago) was a time in which new life forms were appearing and changing rapidly and is often called the *Cambrian Explosion.*

The earliest life forms were marine organisms—algae and other one-celled organisms that either lived freely by themselves or might gather together to form colonies. These early life forms were all soft-bodied, meaning that they did not have hard exoskeletons to protect them. This also means that they did not readily fossilize, and the evidence we find of them often consists of trace fossils and impressions that their bodies left in the sediments.

About 600 million years ago (still in the Precambrian Era), microorganisms first began to develop mineralized exoskeletons for protection. This development was a boon to micro-paleontologists, because it means that these microorganisms were much more likely to fossilize and their fossils are generally better preserved.

The study of micropaleontology is useful to the average person, as well as to the scientist. How so? you ask. Petroleum companies employ microbiologists to study the microfossils in areas where they think there may be large deposits of petroleum (crude oil used to make gasoline, natural gas, kerosene, asphalt and many other derivatives). Micropaleontologists examine the microfossils to determine whether it is likely that the companies will find oil.

Paleontology

Paleontology is the study of prehistoric life, including organisms' evolution and interactions with each other and their environments. As a "historical science" it tries to explain causes rather than conduct experiments to observe effects. Although 5th century BC and medieval thinkers made what would now be called paleontological observations, the science became established in the 18th century as a result of Georges Cuvier's work on comparative anatomy, and developed rapidly in the 19th century. Fossils found in China since the 1990s have provided new information about the earliest evolution of animals, early fish, dinosaurs and the evolution of birds and mammals. Paleontology lies on the border between biology and geology, and also shares with archeology a border that is difficult to define. It now uses techniques drawn from a wide range of sciences, including biochemistry, mathematics and engineering. As knowledge has increased, paleontology has also developed specialised subdivisons, some of which focus on different types of fossil organisms while others focus on ecological and environmental aspects such as ancient climates.

Body fossils and trace fossils are the principal types of evidence about ancient life, and geochemical evidence has helped to decipher the evolution of life before there were organisms large enough to leave fossils. Estimating the dates of these remains is essential but difficult: sometimes adjacent rock layers allow radiometric dating, which provide absolute dates that are accurate to within 0.5%, but more often paleontologists have to rely on relative dating by solving the "jigsaw puzzles" of biostratigraphy. Classifying ancient organisms is also difficult, as many do not fit well into the Linnean taxonomy that is commonly used for classifying living organisms, and paleontologists more often use cladistics to draw up evolutionary "family trees". The final quarter of the 20th century saw the development of molecular phylogenetics, which investigates how closely organisms are related by measuring how similar the DNA is in their genomes. Molecular phylogenetics has also been used to estimate the dates when species diverged, but there is controversy about the reliability of the molecular clock on which such estimates depend.

Use of all these techniques has enabled paleontologists to discover much of the evolutionary history of life, almost all the way back to when Earth became capable of supporting life, about 3,800 million years ago. For about half of that time the only life was single-celled micro-organisms, mostly in microbial mats that formed ecosystems only a few millimeters thick. Earth's atmosphere originally contained virtually no oxygen, and its oxygenation began about 2,400 million years ago. This may have caused an accelerating increase in the diversity and complexity of life, and early multicellular plants and fungi have been found in rocks dated from 1,700 to 1,200 million years ago. The earliest multicellular animal fossils are much later, from about 580 million years ago, but animals diversified very rapidly and there is a lively debate about whether most of this happened in a relatively short Cambrian explosion or started earlier but has been hidden by lack of fossils. All of these organisms lived in water, but plants and invertebrates started colonizing land from about 490 million years ago and vertebrates followed them about 370 million years ago. The first dinosaurs appeared about 230 million years ago and birds evolved from one dinosaur group about 150 million years ago. During the time of the dinosaurs, mammals' ancestors survived only as small, mainly nocturnal insectivores, but after the non-avian dinosaurs became extinct in the Cretaceous–Tertiary extinction event 65 million years ago mammals diversified rapidly. Flowering plants appeared and rapidly diversified between 130 million years ago and 90 million years ago, possibly helped by coevolution with pollinating insects. Social insects appeared around the same time and, although they have relatively few species, now form over 50% of the total mass of all insects. Humans evolved from a lineage of upright-walking apes that appeared 6 to 7 million years ago, and anatomically modern humans appeared under 200,000 years ago. The course of evolution has been changed several times by mass extinctions that wiped out previously dominant groups and allowed other to rise from obscurity to become major components of ecosystems.

Definition

The simplest definition is "the study of ancient life" Paleontology seeks information about several aspects of past organisms: "their identity and origin, their environment and evolution, and what they can tell us about the Earth's organic and inorganic past".

A Historical Science

Paleontology is one of the "historical sciences", along with archaeology, geology, biology, astronomy, cosmogony, philology and history itself This means that it aims to describe phenomena of the past and reconstruct their causes. Hence it has three main elements: description of the phenomena; developing a general theory about the causes of various types of change; and applying those theories to specific facts.

When trying to explain past phenomena, paleontologists and other historical scientists often construct a set of hypotheses about the causes and then look for a "smoking gun", a piece of evidence which indicates that one of the hypotheses is a better explanation than the others. Sometimes the "smoking gun" is discovered by a fortunate accident during other research, for example the discovery by Luis Alvarez and Walter Alvarez of an iridium-rich layer at the Cretaceous-Tertiary boundary made asteroid impact and volcanism the most favoured explanations for the Cretaceous-Tertiary extinction event.

The other main type of science is experimental science, which is often said to work by conducting experiments to *disprove* hypotheses about the workings and causes of natural phenomena – note that this approach cannot prove a hypothesis is correct, since some later experiment may disprove it. However, when confronted with totally unexpected phenomena, such as the first evidence for invisible radiation, experimental scientists often use the same approach as historical scientists: construct a set of hypotheses about the causes and then look for a "smoking gun".

Related Sciences

Paleontology lies on the boundary between biology and geology since paleontology focuses not on life but on fossils, its main source of evidence found in rocks. For historical reasons paleontology is part of the geology departments of many universities, because in the 19th and early 20th centuries geology departments found paleontological evidence important for estimating the ages of rocks while biology departments showed little interest.

Paleontology also has some overlap with archaeology, which primarily works with objects made by humans and with human remains, while paleontologists are interested in the characteristics and evolution of humans as organisms. When dealing with evidence about humans, archaeologists and paleontologists may work together – for example paleontologists might identify animal or plant fossils around an archaeological site, to discover what the people who lived there ate; or they might analyze the climate at the time when the site was inhabited by humans.

In addition paleontology often uses techniques derived from other sciences, including biology, ecology, chemistry, physics and mathematics. For example geochemical signatures from rocks may help to discover when life first arose on Earth, and analyses of carbon isotope ratios may help to identify climate changes and even to explain major transitions such as the

Permian-Triassic extinction event. A relatively recent discipline, molecular phylogenetics, often helps by using comparisons of different modern organisms' DNA and RNA to re-construct evolutionary "family trees"; it has also been used to estimate the dates of important evolutionary developments, although this approach is controversial because of doubts about the reliability of the "molecular clock". Techniques developed in engineering have been used to analyse how ancient organisms might have worked, for example how fast *Tyrannosaurus* could move and how powerful its bite was.

Paleontology even contributes to astrobiology, the investigation of possible life on other planets, by developing models of how life may have arisen and by providing techniques for detecting evidence of life.

Subdivisions

As knowledge has increased, paleontology has developed specialised subdivisons. Vertebrate paleontology concentrates on fossils of vertebrates, from the earliest fish to the immediate ancestors of modern mammals. Invertebrate paleontology deals with fossils of invertebrates such as molluscs, arthropods, annelid worms and echinoderms. Paleobotany focuses on the study of fossil plants, but traditionally includes the study of fossil algae and fungi. Palynology, the study of pollen and spores produced by land plants and protists, straddles the border between paleontology and botany, as it deals with both living and fossil organisms. Micropaleontology deals with all microscopic fossil organisms, regardless of the group to which they belong.

Instead of focusing on individual organisms, paleoecology examines the interactions between different organisms, for example their places in food chains, and the two-way interaction between organisms and their environment – for example the development of oxygenic photosynthesis by bacteria hugely increased the productivity and diversity of ecosystems, and also caused the oxygenation of the atmosphere, which in turn was a prerequisite for the evolution of the most complex eucaryotic cells, from which all multicellular organisms are built

Paleoclimatology, although sometimes treated as part of paleoecology focuses more on the history of Earth's climate and the mechanisms which have changed it – which have sometimes included evolutionary developments, for example the rapid expansion of land plants in the Devonian period removed more carbon dioxide from the atmosphere, reducing the greenhouse effect and thus causing an ice age in the Carboniferous period.

Biostratigraphy, the use of fossils to work out the chronological order in which rocks were formed, is useful to both paleontologists and geologists. Biogeography studies the spatial distribution of organisms, and is also linked to geology, which explains how Earth's geography has changed over time.

Body Fossils

Sources of Evidence

Fossils of organisms' bodies are usually the most informative type of evidence. The most common types are wood, bones, and shells. Fossilisation is a rare event, and most fossils are destroyed by erosion or metamorphism before they can be observed. Hence the fossil record is very incomplete, increasingly so further back in time. Despite this, it is often adequate to illustrate the broader patterns of life's history. There are also biases in the fossil record: different environments are more favourable to the preservation of different types of organism or parts of organisms. Further, only the parts of organisms that were already mineralised are usually preserved, such as the shells of molluscs. Since most animal species are soft-bodied, they decay before they can become fossilised. As a result, although there are 30-plus phyla of living animals, two-thirds have never been found as fossils.

Occasionally, unusual environments may preserve soft tissues. These lagerstätten allow paleontologists to examine the internal anatomy of animals that in other sediments are only represented by shells, spines, claws, etc. – if they are preserved at all. However, even lagerstätten present an incomplete picture of life at the time. The majority of organisms living at the time are probably not represented because lagerstätten are restricted to a narrow range of environments, e.g. where soft-bodied organisms can be preserved very quickly by events such as mudslides; and the exceptional events that cause quick burial make it difficult to study the normal environments of the animals. The sparseness of the fossil record means that organisms are expected to exist long before and after they are found in the fossil record - this is known as the Signor-Lipps effect.

Trace Fossils

Trace fossils consist mainly of tracks and burrows, but also include coprolites (fossil feces) and marks left by feeding. Trace fossils are particularly significant because they represent a data source that is not limited to animals with easily-fossilized hard parts, and which reflects organisms' behaviour. Also many traces date from significantly earlier than the body fossils of animals that are thought to have been capable of making them.[30] Whilst exact assignment of trace fossils to their makers is generally impossible, traces may for example provide the earliest physical evidence of the appearance of moderately complex animals (comparable to earthworms).

Geochemical Observations

Geochemical observations may help to deduce the global level of biological activity, or the affinity of a certain fossil. For example geochemical features of rocks may reveal when life first arose on Earth and may provide evidence of the presence of eucaryotic cells, the type from which all multicellular organisms are built. Analyses of carbon isotope ratios may help to explain major transitions such as the Permian–Triassic extinction event.

Classifying Ancient Organisms

Naming groups of organisms in a way that is clear and widely agreed is important, as some disputes paleontology have just been based on misunderstandings over names. Linnean taxonomy is commonly used for classifying living organisms, but runs into difficulties when dealing with newly-discovered organisms that are significantly different from known ones. For example: it is hard to decide at what level to place a higher-level grouping, e.g. genus or family or order; the Linnean rules for naming groups are tied to their levels, and hence if a group is moved to a different level it has to be renamed.

Paleontologists generally use approaches based on cladistics, a technique for working out the evolutionary "family tree" of a set of organisms It works by the logic that, if groups B and C have more similarities to each other than either has to group A, then B and C are more closely related to each other than either is to A. Characters which are compared may be anatomical, such as the presence of a notochord, or molecular, by comparing sequences of DNA or protein. The result of a successful analysis is a hierarchy of clades – groups whose members are believed to share a common ancestor. Ideally the "family tree" has only two branches leading from each node ("junction"), but sometimes there is insufficient information to achieve this and paleontologists have to make do with junctions that have several branches. The cladistic technique is sometimes fallible, as some features, such as wings or camera eyes, evolved more than once, convergently – this must be taken into account in analyses.

Evolutionary developmental biology, commonly abbreviated to "Evo Devo", also helps paleontologists to produce "family trees", for example the embryological development of some modern brachiopods suggests that brachiopods may be descendants of the halkieriids, which became extinct in the Cambrian period.

Estimating the Dates of Organisms

Paleontology seeks to map out how the character of life has changed through time. A substantial hurdle to this aim is the difficulty of working out how old fossils are. Beds which preserve fossils typically lack radioactive elements on which radiometric dating techniques can be used. This technique is our only means of giving rocks greater than about 50 million years old an absolute age, and can be accurate to within 0.5% or better. Although radiometric dating requires very careful laboratory work, its basic principle is simple: the rates at which various radioactive elements decay are known, and so the ratio of the radioactive element to the element into which it decays can be used to calculate how long has passed since the radioactive element was incorporated into the rock. Radioactive elements are only common in rocks with a volcanic origin, and the only fossil-bearing rocks that can be dated radiometrically are a few volcanic ash layers

In general, palaeontologists must rely on stratigraphy to date fossils. Stratigraphy is the science of deciphering the "layer-cake" that is the sedimentary record, and has been compared to a jigsaw puzzle. Rocks normally form relatively flat-lying layers, with each layer younger than the one underneath it. If a fossil is found between two layers whose ages are known, the fossil's age must be between the two known ages. Because rock sequences are not continuous, but may be broken up by faults or periods of erosion, it is very difficult to correlate between rock beds which are not directly next to one another. Fossils of species that survived for a relatively short time can be used to link up isolated rocks. For instance, the conodont *Eoplacognathus pseudoplanus* has a short range in the Middle Ordovician period rocks of unknown age are found to have traces of *E. pseudoplanus*, they must have a mid-Ordovician age. There are some difficulties with biostratigraphy, though - chosen index fossils must be distinctive, globally distributed and have a short time range to be useful. They may also turn out to have longer fossil ranges than first thought. Stratigraphy and biostratigraphy can in general provide only relative dating (*A* was before *B*), which is often sufficient for studying processes of evolution. However, this is difficult for some time periods, because of the problems involved in matching up rocks of the same age across different continents.

Family tree relationships may also help to narrow down the date on which lineages first appeared. For instance, if fossils of B or C date to X million years ago and the calculated "family tree" says A was an ancestor of B and C, then A must have evolved more than X million years ago.

It is also possible to estimate how long ago two living clades diverged – i.e. approximately how long ago their last common ancestor must have lived – by assuming that DNA mutations accumulate at a constant rate. These "molecular clocks", however, are fallible, and provide only a very approximate timing: for example they are not sufficiently precise and reliable for estimating when the groups that feature in the Cambrian explosion first evolved and estimates produced by different techniques may vary by a factor of two.

HISTORY OF LIFE

The evolutionary history of life stretches back to over 3000 million years ago, possibly as far as 3800 million years ago. Earth formed about 4540 million years ago and, after a collision that formed the Moon about 40 million years later, may have cooled quickly enough to have oceans and an atmosphere about 4440 million years ago However there is evidence on the Moon of a Late Heavy Bombardment from 4000 to 3800 million years ago. If, as seem likely, such a bombardment struck Earth at the same time, the first atmosphere and oceans may have been stripped away. The oldest undisputed evidence of life on Earth dates to 3000 million years ago, although there have been reports, often disputed, of fossil bacteria from 3400 million years ago and of geochemical evidence for the presence of life 3800 million years ago. Even the simplest modern organisms are too complex to have emerged directly from non-living materials. Some scientists have proposed that life on Earth was "seeded" from elsewhere, but most research concentrates on various explanations of how life could have arisen independently on Earth.

For about 2000 million years microbial mats, multi-layered colonies of different types of bacteria, were the dominant life on Earth. The evolution of oxygenic photosynthesis enabled them to play the major role in the oxygenation of the atmosphere for which there is geological evidence from about 2400 million years ago, and increased their effectiveness as nurseries of evolution While eucaryotes, cells with complex internal structures, may have been present earlier, their evolution speeded up when they acquired the ability to transform oxygen from a poison to a powerful "fuel" for their metabolisms, a development that may have started with their capturing oxygen-powered bacteria as endosymbionts. The earliest evidence of complex eucaryotes with organelles, organs within a cell, dates from 1,850 million years ago.

Multicellular life is composed only of eucaryotic cells, and the earliest evidence for it is from 1700 million years ago, although specialization of cells for different functions first appears between 1430 million years ago (a possible fungus) and 1200 million years ago (a probable red alga). Sexual reproduction may be a prerequisite for specialization of cells, as an asexual multicellular organism would be at risk of being taken over by rogue cells which retain the ability to reproduce.

The earliest known animals are cnidarians from about 580 million years ago, but these are so modern-looking that the earliest animals must have appeared before then. Early fossils of animals are rare because they did not develop mineralized hard parts that fossilize easily until about 548 million years ago The earliest modern-looking bilaterian animals appear in the Early Cambrian, along with several "weird wonders" that bear little obvious resemblance to any modern animals. There is a long-running debate about whether this Cambrian explosion was truly a very rapid period of evolutionary experimentation; alternative views are that modern-looking animals began evolving earlier but fossils of their precursors have not yet been found, or that the "weird wonders" are evolutionary "aunts" and "cousins" of modern groups Vertebrates remained an obscure group until the first fish with jaws appeared in the Late Ordovician.

The spread of life from water to land required organisms to solve several problems, including protection against drying out and supporting themselves against gravity The earliest evidence of land plants and land invertebrates date back to about 476 million years ago and 490 million years ago respectively The lineage that produced land vertebrates evolved later but very rapidly between 370 million years ago and 360 million years ago; recent discoveries have overturned earlier ideas about the history and driving forces behind their evolution Land plants were so successful that they caused an ecological crisis in the Late Devonian, until the evolution and spread of fungi that could digest dead wood.

During the Permian period synapsids, including the ancestors of mammals, may have dominated land environments but the Permian–Triassic extinction event 251 million years ago came very close to wiping out complex life During the slow recovery from this catastrophe a previously obscure group, archosaurs, became the most abundant and diverse terrestrial vertebrates. One archosaur group, the dinosaurs, were the dominant land vertebrates for the rest of the Mesozoic and birds evolved from one group of dinosaurs. During this time mammals' ancestors survived only as small, mainly nocturnal insectivores, but this apparent set-back may have accelerated the development of mammalian traits such as endothermy and hair. After the Cretaceous–Tertiary extinction event 65 million years ago killed off the non-avian dinosaurs – birds are the only surviving dinosaurs – mammals increased rapidly in size and diversity, and some took to the air and the sea.

Fossil evidence indicates that flowering plants appeared and rapidly diversified in the Early Cretaceous, between 130 million years ago and 90 million years ago Their rapid rise to dominance of terrestrial ecosystems is thought to have been propelled by coevolution with pollinating insects. Social insects appeared around the same time and, although they account for only small parts of the insect "family tree", now form over 50% of the total mass of all insects.

Humans evolved from a lineage of upright-walking apes whose earliest fossils date from over six million years ago. Although early members of this lineage had chimp-sized brains, about 25% as big as modern humans', there are signs of a steady increase in brain size after about three million years ago. There is a long-running debate about whether *modern* humans are descendants of a single small population in Africa, which then migrated all over the world less than 200,000 years ago and replaced previous hominine species, or arose.

MASS EXTINCTIONS

Life on earth has suffered occasional mass extinctions at least since 542 million years ago. Although they are disasters at the time, mass extinctions have sometimes accelerated the evolution of life on earth. When dominance of particular ecological niches passes from one group of organisms to another, it is rarely because the new dominant group is "superior" to the old and usually because an extinction event eliminates the old dominant group and makes way for the new one.

The fossil record appears to show that the rate of extinction is slowing down, with both the gaps between mass extinctions becoming longer and the average and background rates of extinction decreasing. However, it is not certain if the actual rate of extinction has altered, since both of these observations could be explained in several ways:

- The oceans may have become more hospitable to life over the past 500 million years and less vulnerable to mass extinctions dissolved oxygen became more widespread and penetrated to greater depths; the development of life on land reduced the run-off of nutrients and hence the risk of eutrophication and anoxic events; and marine ecosystems became more diversified so that food chains were less likely to be disrupted.
- Reasonably complete fossils are very rare, most extinct organisms are represented only by partial fossils, and complete fossils are rarest in the oldest rocks. So paleontologists have mistakenly assigned parts of the same organism to different genera which were often defined solely to accommodate these finds – the story of *Anomalocaris* is an example of this. The risk of this mistake is higher for older fossils because these are often unlike parts of any living organism. Many of the "superfluous" genera are represented by fragments which are not found again and the "superfluous" genera appear to become extinct very quickly.

Biodiversity in the fossil record, which is "the number of distinct genera alive at any given time; that is, those whose first occurrence predates and whose last occurrence postdates that time" shows a different trend a fairly swift rise from 542 to 400 million years ago; a slight decline from 400 to 200 million years ago, in which the devastating Permian–Triassic extinction event is an important factor; and a swift rise from 200 million years ago to the present.

History of Paleontology

Although palaeontology become established around 1800, earlier thinkers had noticed aspects of the fossil record. The ancient Greek philosopher Xenophanes (570-480 BC) wrote about fossil sea shells indicating that land was once under water. During the Middle Ages the Persian naturalist Ibn Sina, (known as *Avicenna* in Europe, discussed fossils and proposed a theory of petrifying fluids that Albert of Saxony elaborated on in the 14th century The Chinese naturalist Shen Kuo (1031-1095) proposed a theory of climate change based on the presence of petrified bamboo in regions that in his time were too dry for bamboo.

In early modern Europe, the systematic study of fossils emerged as an integral part of the changes in natural philosophy that occurred during the Age of Reason. At the end of the 18th century Georges Cuvier's work established comparative anatomy as a scientific discipline and, by proving that some fossils animals resembled no living ones, demonstrated that animals could become extinct and led to the emergence of paleontology. The expanding knowledge of the fossil record also played an increasing role in the development of geology, particularly stratigraphy.

The first half of the 19th century saw geological and paleontological activity become increasingly well organized with the growth of geologic societies and museums, and an increasing number of professional geologists and fossil specialists. Interest increased for reasons that were not purely scientific, as geology and paleontology helped industrialists to find and exploit natural resources such as coal.

This contributed to a rapid increase in knowledge about the history of life on Earth and to progress in the definition of the geologic time scale, largely based on fossil evidence. In 1822 the word "paleontology" was invented by the editor of a French scientific journal to refer to the study of ancient living organisms through fossils As knowledge of life's history continued to improve, it became increasingly obvious that there had been some kind of successive order to the development of life. This would encourage early evolutionary theories on the transmutation of species After Charles Darwin published *Origin of Species* in 1859, much of the focus of paleontology shifted to understanding evolutionary paths, including human evolution, and evolutionary theory.

The last half of the 19th century saw a tremendous expansion in paleontological activity, especially in North America. The trend continued in the 20th century with additional regions of the Earth being opened to systematic fossil collection. Fossils found in China near the end of the 20th century have been particularly important as they have provided new information about the earliest evolution of animals, early fish, dinosaurs and the evolution of birds The last few decades of the 20th century saw a renewed interest in mass extinctions and their role in the evolution of life on Earth There was also a renewed interest in the Cambrian explosion that apparently saw the development of the body plans of most animal phyla. The discovery of fossils of the Ediacaran biota and developments in paleobiology extended knowledge about the history of life back far before the Cambrian. Increasing awareness of Gregor Mendel's pioneering work in genetics led first to the development of population genetics and then in the mid-20th century to the modern evolutionary synthesis, which explains evolution as the outcome of events such as mutations and horizontal gene transfer providing genetic variation, with genetic drift and natural selection driving changes in this variation over time A few years later Watson and Crick discovered the structure of DNA and confirmed its role in genetic inheritance, which is now known as the "Central Dogma" of molecular biology. In the 1960s molecular phylogenetics, the investigation of evolutionary "family trees" by biochemistry techniques, began to make an impact, particularly when it was proposed that the human lineage had diverged from apes much more recently than was generally thought at the time. Although this early study compared proteins from apes and humans, most molecular phylogenetics research is now based on comparisons of RNA and DNA.

CHAPTER–2
Fossils

INTRODUCTION

Fossils (from Latin *fossus*, literally "having been dug up") are the preserved remains or traces of animals, plants, and other organisms from the remote past. The totality of fossils, both discovered and undiscovered, and their placement in fossiliferous (fossil-containing) rock formations and sedimentary layers (strata) is known as the *fossil record*. The study of fossils across geological time, how they were formed, and the evolutionary relationships between taxa (phylogeny) are some of the most important functions of the science of paleontology.

Fossils are typically distinguished by minimum age, most often the arbitrary date of 10,000 years ago. Hence, fossils range in age from the youngest at the start of the Holocene Epoch to the oldest from the Archaean Eon several billion years old. The observations that certain fossils were associated with certain rock strata led early geologists to recognize a geological timescale in the 19th century. The development of radiometric dating techniques in the early 20th century allowed geologists to determine the numerical or *"absolute" age* of the various strata and thereby the included fossils.

Like extant organisms, fossils vary in size from microscopic, such as single bacterial cells only one micrometer in diameter, to gigantic, such as dinosaurs and trees many meters long and weighing many tons. A fossil normally preserves only a portion of the deceased organism, usually that portion that was partially mineralized during life, such as the bones and teeth of vertebrates, or the chitinous exoskeletons of invertebrates. Preservation of soft tissues is exquisitely rare in the fossil record. Fossils may also consist of the marks left behind by the organism while it was alive, such as the footprint or feces (coprolites) of a reptile. These types of fossil are called trace fossils (or *ichnofossils*), as opposed to *body fossils*. Finally, past life leaves some markers that cannot be seen but can be detected in the form of biochemical signals; these are known as *chemofossils* or *biomarkers*.

PLACES OF EXCEPTIONAL FOSSILIZATION

Fossil sites with exceptional preservation — sometimes including preserved soft tissues — are known as Lagerstätten. These formations may have resulted from carcass burial in an anoxic environment with minimal bacteria, thus delaying decomposition. Lagerstätten span geological time from the Cambrian period to the present. Worldwide, some of the best examples of near-perfect fossilization are the Cambrian Maotianshan shales and Burgess Shale, the Devonian Hunsrück Slates, the Jurassic Solnhofen limestone, and the Carboniferous Mazon Creek localities.

EARLIEST FOSSILIFEROUS SITES

Earth's oldest fossils are the stromatolites consisting of rock built from layer upon layer of sediment and other precipitants Based on studies of now-rare (but living) stromatolites (specifically, certain blue-green bacteria), the growth of fossil stromatolitic structures was biogenetically mediated by mats of microorganisms through their entrapment of sediments. However, abiotic mechanisms for stromatolitic growth are also known, leading to a decades-long and sometimes-contentious scientific debate regarding biogenesis of certain formations, especially those from the lower to middle Archaean eon.

It is most widely accepted that stromatolites from the late Archaean and through the middle Proterozoic eon were mostly formed by massive colonies of cyanobacteria (formerly known as blue-green "algae"), and that the oxygen byproduct of their photosynthetic metabolism first resulted in earth's massive banded iron formations and subsequently oxygenated earth's atmosphere.

Even though it is extra rare, microstructures resembling cells are sometimes found within stromatolites; but these are also the source of scientific contention. The Gunflint Chert contains abundant microfossils widely accepted as a diverse consortium of 2.0 bya microbes

In contrast, putative fossil cyanobacteria cells from the 3.4 bya Warrawoona Group in Western Australia are in dispute since abiotic processes cannot be ruled out. Confirmation of the Warrawoona microstructures as cyanobacteria would profoundly impact our understanding of when and how early life diversified, pushing important evolutionary milestones further back in time (reference). The continued study of these oldest fossils is paramount to calibrate complementary molecular phylogenetics models.

DEVELOPMENTS IN INTERPRETATION OF THE FOSSIL RECORD

Ever since recorded history began, and probably before, people have noticed and gathered fossils, including pieces of rock and minerals that have replaced the remains of biologic organisms, or preserved their external form. Fossils themselves, and the totality of their occurrence within the sequence of Earth's rock strata is referred to as the fossil record.

The fossil record was one of the early sources of data relevant to the study of evolution and continues to be relevant to the history of life on Earth. Paleontologists examine the fossil record in order to understand the process of evolution and the way particular species have evolved.

Explanations

Various explanations have been put forth throughout history to explain what fossils are and how they came to be where they were found. Many of these explanations relied on folktales or mythologies. In China the fossil bones of ancient mammals including *Homo erectus* were often mistaken for "dragon bones" and used as medicine and aphrodisiacs. In the West the presence of fossilized sea creatures high up on mountainsides was seen as proof of the biblical deluge.

Greek scholar Aristotle realized that fossil seashells from rocks were similar to those found on the beach, indicating the fossils were once living animals. Leonardo da Vinci concurred with Aristotle's view that fossils were the remains of ancient life in 1027, the Persian geologist, Ibn Sina (known as *Avicenna* in Europe), explained how the stoniness of fossils was caused in *The Book of Healing*. However, he rejected the explanation of fossils as organic remains Aristotle previously explained it in terms of vapourous exhalations, which Ibn Sina modified into the theory of petrifying fluids (*Succus lapidificatus*), which was elaborated on by Albert of Saxony in the 14th century and accepted in some form by most naturalists by the 16th century. Ibn Sina gave the following explanation for the origin of fossils from the petrifaction of plants and animals:

> "If what is said concerning the petrifaction of animals and plants is true, the cause of this (phenomenon) is a powerful mineralizing and petrifying virtue which arises in certain stony spots, or emanates suddenly from the earth during earthquake and subsidences, and petrifies whatever comes into contact with it. As a matter of fact, the petrifaction of the bodies of plants and animals is not more extraordinary than the transformation of waters."

More scientific views of fossils emerged during the Renaissance. For example, Leonardo Da Vinci noticed discrepancies with the use of the biblical flood narrative as an explanation for fossil origins:

> "If the Deluge had carried the shells for distances of three and four hundred miles from the sea it would have carried them mixed with various other natural objects all heaped up together; but even at such distances from the sea we see the oysters all together and also the shellfish and the cuttlefish and all the other shells which congregate together,

found all together dead; and the solitary shells are found apart from one another as we see them every day on the sea-shores.

And we find oysters together in very large families, among which some may be seen with their shells still joined together, indicating that they were left there by the sea and that they were still living when the strait of Gibraltar was cut through. In the mountains of Parma and Piacenza multitudes of shells and corals with holes may be seen still sticking to the rocks..."

William Smith (1769-1839), an English canal engineer, observed that rocks of different ages (based on the law of superposition) preserved different assemblages of fossils, and that these assemblages succeeded one another in a regular and determinable order. He observed that rocks from distant locations could be correlated based on the fossils they contained. He termed this the principle of faunal succession.

Smith, who preceded Charles Darwin, was unaware of biological evolution and did not know why faunal succession occurred. Biological evolution explains why faunal succession exists: as different organisms evolve, change and go extinct, they leave behind fossils. Faunal succession was one of the chief pieces of evidence cited by Darwin that biological evolution had occurred.

Georges Cuvier came to believe that most if not all the animal fossils he examined were remains of species that were now extinct. This led Cuvier to become an active proponent of the geological school of thought called catastrophism. Near the end of his 1796 paper on living and fossil elephants he said:

> All of these facts, consistent among themselves, and not opposed by any report, seem to me to prove the existence of a world previous to ours, destroyed by some kind of catastrophe.

Biological Explanations

Early naturalists well understood the similarities and differences of living species leading Linnaeus to develop a hierarchical classification system still in use today. It was Darwin and his contemporaries who first linked the hierarchical structure of the great tree of life in living organisms with the then very sparse fossil record. Darwin eloquently described a process of descent with modification, or evolution, whereby organisms either adapt to natural and changing environmental pressures, or they perish.

When Charles Darwin wrote *On the Origin of Species by Means of Natural Selection, or the Preservation of Favoured Races in the Struggle for Life*, the oldest animal fossils were those from the Cambrian Period, now known to be about 540 million years old. The absence of older fossils worried Darwin about the implications for the validity of his theories, but he expressed hope that such fossils would be found, noting that: "only a small portion of the world is known with accuracy." Darwin also pondered the sudden appearance of many groups (i.e. phyla) in the oldest known Cambrian fossiliferous strata.

Further Discoveries

Since Darwin's time, the fossil record has been pushed back to between 2.3 and 3.5 billion years before the present. Most of these Precambrian fossils are microscopic bacteria or microfossils. However, macroscopic fossils are now known from the late Proterozoic. The Ediacaran biota (also called Vendian biota) dating from 575 million years ago collectively constitutes a richly diverse assembly of early multicellular eukaryotes.

The fossil record and faunal succession form the basis of the science of biostratigraphy or determining the age of rocks based on the fossils they contain. For the first 150 years of geology, biostratigraphy and superposition were the only means for determining the relative age of rocks. The geologic time scale was developed based on the relative ages of rock strata as determined by the early paleontologists and stratigraphers.

Since the early years of the twentieth century, absolute dating methods, such as radiometric dating (including potassium/argon, argon/argon, uranium series, and carbon-14 dating) have been used to verify the relative ages obtained by fossils and to provide absolute ages for many fossils. Radiometric dating has shown that the earliest known stromatolites are over 3.4 billion years old. Various dating methods have been used and are used today depending on local geology and context, and while there is some variance in the results from these dating methods, nearly all of them provide evidence for a very old Earth, approximately 4.6 billion years.

Modern View

"The fossil record is life's evolutionary epic that unfolded over four billion years as environmental conditions and genetic potential interacted in accordance with natural selection." The earth's climate, tectonics, atmosphere, oceans, and periodic disasters invoked the primary selective pressures on all organisms, which they either adapted to, or they perished with or without leaving descendants. Modern paleontology has joined with evolutionary biology to share the interdisciplinary task of unfolding the tree of life, which inevitably leads backwards in time to the microscopic life of the Precambrian when cell structure and functions evolved. Earth's deep time in the Proterozoic and deeper still in the Archaean is only "recounted by microscopic fossils and subtle chemical signals." Molecular biologists, using phylogenetics, can compare protein amino acid or nucleotide sequence homology (i.e., similarity) to infer taxonomy and evolutionary distances among organisms, but with limited statistical confidence. The study of fossils, on the other hand, can more specifically pinpoint when and in what organism branching occurred in the tree of life. Modern phylogenetics and paleontology work together in the clarification of science's still dim view of the appearance of life and its evolution during deep time on earth.

Niles Eldredge's study of the *Phacops* trilobite genus supported the hypothesis that modifications to the arrangement of the trilobite's eye lenses proceeded by fits and starts over millions of years during the Devonian Eldredge's interpretation of the *Phacops* fossil record was that the aftermaths of the lens changes, but not the rapidly occurring evolutionary process, were fossilized. This and other data led Stephen Jay Gould and Niles Eldredge to publish the seminal paper on punctuated equilibrium in 1971.

Examples of Modern Development

An example of modern paleontological progress is the application of synchrotron X-ray tomographic techniques to early Cambrian bilaterian embryonic microfossils that has recently yielded new insights of metazoan evolution at its earliest stages. The tomography technique provides previously unattainable three-dimensional resolution at the limits of fossilization. Fossils of two enigmatic bilaterians, the worm-like *Markuelia* and a putative, primitive protostome, *Pseudooides,* provide a peek at germ layer embryonic development. These 543-million-year-old embryos support the emergence of some aspects of arthropod development earlier than previously thought in the late Proterozoic. The preserved embryos from China and Siberia underwent rapid diagenetic phosphatization resulting in exquisite preservation, including cell structures. This research is a notable example of how knowledge encoded by the fossil record continues to contribute otherwise unattainable information on the emergence and development of life on Earth. For example, the research suggests *Markuelia* has closest affinity to priapulid worms, and is adjacent to the evolutionary branching of Priapulida, Nematoda and Arthropoda.

RARITY OF FOSSILS

Fossilization is an exceptionally rare occurrence, because most components of formerly-living things tend to decompose relatively quickly following death. In order for an organism to be fossilized, the remains normally need to be covered by sediment as soon as possible. However there are exceptions to this, such as if an organism becomes frozen, desiccated, or comes to rest in an anoxic (oxygen-free) environment. There are several different types of fossils and fossilization processes.

Due to the combined effect of taphonomic processes and simple mathematical chance, fossilization tends to favor organisms with hard body parts, those that were widespread, and those that lived for a long time. On the other hand, it is very unusual to find fossils of small, soft bodied, geographically restricted and geologically ephemeral organisms, because of their relative rarity and low likelihood of preservation.

Larger specimens (macrofossils) are more often observed, dug up and displayed, although microscopic remains (microfossils) are actually far more common in the fossil record.

Some casual observers have been perplexed by the rarity of transitional species within the fossil record. The conventional explanation for this rarity was given by Darwin, who stated that "the extreme imperfection of the geological record," combined with the short duration and narrow geographical range of transitional species, made it unlikely that many such fossils would be found. Simply put, the conditions under which fossilization takes place are quite rare; and it is highly unlikely that any given organism will leave behind a fossil. Eldredge and Gould developed their theory of punctuated equilibrium in part to explain the pattern of stasis and sudden appearance in the fossil record.

Types of Preservation

Permineralization

Permineralization occurs after burial, as the empty spaces within an organism (spaces filled with liquid or gas during life) become filled with mineral-rich groundwater and the minerals precipitate from the groundwater, thus occupying the empty spaces. This process can occur in very small spaces, such as within the cell wall of a plant cell. Small scale permineralization can produce very detailed fossils. For permineralization to occur, the organism must become covered by sediment soon after death or soon after the initial decaying process. The degree to which the remains are decayed when covered determines the later details of the fossil. Some fossils consist only of skeletal remains or teeth; other fossils contain traces of skin, feathers or even soft tissues. This is a form of diagenesis.

Casts and Molds

In some cases the original remains of the organism have been completely dissolved or otherwise destroyed. When all that is left is an organism-shaped hole in the rock, it is called an *external mold*. If this hole is later filled with other minerals, it is a *cast*. An *internal mold* is formed when sediments or minerals fill the internal cavity of an organism, such as the inside of a bivalve or snail.

Replacement and Recrystallization

Replacement occurs when the shell, bone or other tissue is replaced with another mineral. In some cases mineral replacement of the original shell occurs so gradually and at such fine scales that microstructural features are preserved despite the total loss of original material. A shell is said to be *recrystallized* when the original skeletal minerals are still present but in a different crystal form, as from aragonite to calcite.

Compression Fossils

Compression fossils, such as those of fossil ferns, are the result of chemical reduction of the complex organic molecules composing the organism's tissues. In this case the fossil consists of original material, albeit in a geochemically altered state. Often what remains is a carbonaceous film. This chemical change is an expression of diagenesis.

Bioimmuration

The star-shaped holes (*Catellocaula vallata*) in this Upper Ordovician bryozoan represent a soft-bodied organism preserved by bioimmuration in the bryozoan skeleton

Bioimmuration is a type of preservation in which a skeletal organism overgrows or otherwise subsumes another organism, preserving the latter, or an impression of it, within the skeleton. Usually it is a sessile skeletal organism, such as a bryozoan or an oyster, which grows along a substrate, covering other sessile encrusters. Sometimes the bioimmured organism is soft-bodied and is then preserved in negative relief as a kind of external mold. There are also cases where an organism settles on top of a living skeletal organism which grows upwards, preserving the settler in its skeleton. Bioimmuration is known in the fossil record from the Ordovician to the Recent.

To sum up, fossilization processes proceed differently for different kinds of tissues and under different kinds of conditions.

TRACE FOSSILS

Microfossils

'Microfossil' is a descriptive term applied to fossilized plants and animals whose size is just at or below the level at which the fossil can be analyzed by the naked eye. A commonly applied cut-off point between "micro" and "macro" fossils is 1 mm, although this is only an approximate guide. Microfossils may either be complete (or near-complete) organisms in themselves (such as the marine plankters foraminifera and coccolithophores) or component parts (such as small teeth or spores) of larger animals or plants. Microfossils are of critical importance as a reservoir of paleoclimate information, and are also commonly used by biostratigraphers to assist in the correlation of rock units.

RESIN FOSSILS

Fossil resin (colloquially called amber) is a natural polymer found in many types of strata throughout the world, even the Arctic. The oldest fossil resin dates to the Triassic, though most dates to the Tertiary. The excretion of the resin by certain plants is thought to be an evolutionary adaptation for protection from insects and to seal wounds caused by damage elements. Fossil resin often contains other fossils called inclusions that were captured by the sticky resin. These include bacteria, fungi, other plants, and animals. Animal inclusions are usually small invertebrates, predominantly arthropods such as insects and spiders, and only extremely rarely a vertebrate such as a small lizard. Preservation of inclusions can be exquisite, including small fragments of DNA.

PSEUDOFOSSILS

Pseudofossils are visual patterns in rocks that are produced by naturally occurring geologic processes rather than biologic processes. They can easily be mistaken for real fossils. Some pseudofossils, such as dendrites, are formed by naturally occurring fissures in the rock that get filled up by percolating minerals. Other types of pseudofossils are kidney ore (round shapes in iron ore) and moss agates, which look like moss or plant leaves. Concretions, spherical or ovoid-shaped nodules found in some sedimentary strata, were once thought to be dinosaur eggs, and are often mistaken for fossils as well.

LIVING FOSSILS

Living fossil is an informal term used for any living species which closely resembles a species known from fossils — that is, it is as if the ancient fossil had "come to life."

This can be (a) a species or taxon known only from fossils until living representatives were discovered, such as the lobe-finned coelacanth, primitive monoplacophoran mollusk, and the Chinese maidenhair tree, or (b) a single living species with no close relatives, such as the New Caledonian Kagu, or the Sunbittern, or (c) a small group of closely-related species with no other close relatives, such as the oxygen-producing, primoidial stromatolite, inarticulate lampshell Lingula, many-chambered pearly *Nautilus*, rootless whisk fern, armored horseshoe crab, and dinosaur-like tuatara that are the sole survivors of a once large and widespread group in the fossil record.

CHAPTER–3

Microfossils

INTRODUCTION

Microfossils are the tiny remains of bacteria, protists, fungi, animals, and plants. Microfossils are a heterogeneous bunch of fossil remains studied as a single discipline because rock samples must be processed in certain ways to remove them and microscopes must be used to study them. Thus, microfossils, unlike other kinds of fossils, are not grouped according to their relationships to one another, but only because of their generally small size and methods of study. For example, fossils of bacteria, foraminifera, diatoms, very small invertebrate shells or skeletons, pollen, and tiny bones and teeth of large vertebrates, among others, can be called microfossils. But it is an unnatural grouping. Nevertheless, this utilitarian subdivision of paleontology, first recognized in 1883, is very significant in geology, paleontology, and biology.

Microfossils are perhaps the most important group of all fossils — they are extremely useful in age-dating, correlation and paleoenvironmental reconstruction, all important in the oil, mining, engineering, and environmental industries, as well as in general geology. Billions of dollars have been made on the basis of microfossil studies. Because they usually occur in huge numbers in all kinds of sedimentary rocks, they are the most abundant and most easily accessible fossils. Indeed, some very thick rock layers are made entirely of microfossils. The pyramids of Egypt are made of sedimentary rocks, for example, that consist of the shells of foraminifera, a major microfossil group.

Microfossils can also be very useful in teaching science at all levels. Students are commonly fascinated by things they cannot see with their naked eyes, especially when the objects are beautiful or interesting in their own right. Furthermore, collection of microfossils is usually possible close to many schools — in fact, some schools are built right on top of microfossil-bearing sedimentary rocks! Processing the rock samples is usually easy and safe enough for children to do themselves, or at least to watch. Prepared samples can be purchased or obtained from museums and some universities. Because so many microfossils are usually found in any sample, the students can even keep their own finds!

Although plants and animals are the most obvious life around us today, they are not the most numerous nor the most important contributors of microfossils. Bacteria (prokaryotes) and protists far outnumber them, live in more diverse habitats, and leave a greater diversity of microfossils. Today these organisms live from Antarctic ice deserts to steaming volcanic hot springs, and from the highest mountains to the deepest sea. Some cause diseases, such as malaria which infects 350-400 million people today; others are useful to humans. Most simply live their lives unknown to us but contributing enormously to our well being through the production of oxygen, the degradation of waste materials, recycling of nutrients, production of food, and a multitude of other functions, some of which take place in our own bodies. Fungi, another group in modern environments that both benefit and plague humans, have a long, but mostly unstudied, microfossil record.

Prokaryotes and protists are very well represented in the fossil record. Prokaryotes are the oldest known fossils, and they were the only life on Earth for most of its history — from 3.5 to 1.5 billion years ago. Protists joined them at least 1.5 bya, and animals and plants were latecomers at less than .55 bya. All these microfossils provide insights to Earth and life history, and so are important to study in paleontology. The single-celled forms help to develop and test evolutionary models using organisms that are not multicellular or sexual in all cases, and with greater ecological variety.

Generally prokaryotes and protists are single-celled. Yet the most significant contrast among life forms separates them. Prokaryotes have their DNA loosely organized within the cell and not in the cell nucleus, and chromosomes are absent. All protists, fungi, animals and plants are eukaryotes and have chromosomes made of DNA, RNA, and proteins in a nucleus. Many other very important differences occur too (Table 3.1). Animals, plants and fungi are multicellular; protists are generally unicellular and include all other eukaryotes.

Table 3.1: Some primary differences between prokaryotes and eukaryotes

PROKARYOTES	EUKARYOTES
Nucleus absent	Nucleus present
Meiosis absent	Meiosis
1 basic genome	Chromosome number 2-600
Mitochondria absent	Mitochondria present
Chloroplasts absent	Chloroplasts may be present
Endoplasmic reticulum absent	Endoplasmic reticulum present
Vacuoles absent	Vacuoles present

Source: Lipps, J.H. (1992), *Fossil Prokaryotes* and *Protists*, 342 p.

Prokaryotes and protists are often called "simple", but this is just not true. Each one must do everything with just a single cell that higher plants or animals do with millions of cells. Single-celled organisms have many different kinds of specialized organelles within their cells that function in extraordinary ways. Prokaryotes, protists, fungi, animals, and plants are all very successful at making a living, and that is all that evolution requires. Although prokaryotes and protists seem simpler, they arose much earlier than their multicellular descendants and so might be considered more primitive, but some have also existed for at least 3.5 billion years and must therefore be considered very successful indeed.

Fungi, plants, and animals contribute a vast multitude of small parts to the microfossil record. Fungi are found as isolated microscopic filaments and spherical spores, usually associated with larger fossil plant material. As such, they have largely been ignored by paleontologists. Many plants have small pieces and parts that can be found as microfossils. Most important of these are pollen and spores which can be very abundant in terrestrial and nearshore marine deposits. Just about any animal with skeletal parts also contributes to the microfossil record.

RECORD OF MICROFOSSILS

Most organisms lack substantial hard parts and rarely fossilize. The fossil record in its entirety is estimated to include only from 3% to 13.6% of all species that ever lived. The vast majority of described species are living — only 8.7% of those described formally are fossils. No one has ever calculated how many of those are microfossils, but probably between 1/4 and 1/2, largely because they are so useful in geologic applications. Most prokaryotes and protists do not have skeletons of any sort, and have no fossil record whatsoever. Nevertheless, remarkable prokaryotes are known from very old Precambrian rocks, and dome-shaped stromatolites (built by photosynthesizing cyanobacteria) are very common from 3.5 to about .4 billion years ago and less common right up to the Recent. Among the described species of protists, about 50% are fossil, again because they are so useful in age dating, correlation and paleoenvironmental analysis. However, entire groups like the ciliates (with the exception of the small group of tintinnids which do have shells) are virtually absent from the fossil record. Parasitic prokaryotes, protists, and animals, have surely inhabited other organisms for billions of years, but have no fossil record of any import whatsoever. Indeed, even among these larger, multicellular groups, the fossil record is nowhere near complete. In spite of these problems, microfossils provide one of the best records of evolution's course because they are generally rather abundant and complete.

SYSTEMATICS AND RELATIONSHIPS AMONG MICROFOSSILS

Because microfossils are an arbitrary grouping based on methods of study, no single classification of them will suffice except at the highest levels. The systematics of prokaryotes and protists has long been confused and multifaceted at all levels. For example, among the prokaryotes, the cyanobacteria were thought to be plants called blue-green algae. We now know that they are really bacteria. Protist classification is very complex. They have traditionally been classified along with animals and plants. "Animal-like" protists, or those that can move on their own, were claimed by protozoologists; "plant-like" protists, or those that generally photosynthesize, were studied by phycologists. This dichotomy produced considerable confusion and impeded understanding of the relationships among the various kinds. The different investigators soon formed their own exclusive classifications, all of which are now under reconsideration based on molecular studies. Likewise, bacteria have not been clearly deciphered, although molecular techniques are helping enormously here as well. Animals, plants and fungi have mostly held their own as inclusive groups, although modern paleontological and molecular studies are changing views within each group as well.

The relationships among life forms are complex. A satisfactory overview of what kinds of life there are on Earth is still ephemeral. The following six domains of life have been proposed:

1. Archaebacteria;
2. Eubacteria;
3. Protista;
4. Animalia;
5. Plantae; and
6. Fungi.

All these groups make microfossils. However, not all are easily studied as microfossils. Many require special techniques of removing them from rocks or making them visible, others require very specialized equipment for their study. But a good number of very important ones are easily studied by any students with a pot to boil the rock sample, a simple microscope, and curiosity about the world near them.

IMPORTANT AND EASILY STUDIED MICROFOSSILS

Not all microfossils can be easily used in classrooms. Bacterial and fungal microfossils are usually difficult to study, unless a prepared thin-section of rock containing them is available. Such thin sections can sometimes be purchased from scientific supply houses. Likewise, pollen and spores can be had in prepared slides. These microfossils are generally very small and require a fairly powerful microscope for viewing. This kind of exercise, then, becomes observational but can be used successfully for interpretation.

Protists include several groups that have shells or skeletons of hard materials that preserve well, are easy to extract from rocks, and can be seen with a classroom dissecting microscope). These are ideal for students of most any age to study. They include foraminifera, radiolaria and diatoms. Of course, other kinds may also occur in the same rocks, but they will be hard to find and difficult to see.

Foraminifera

Foraminifera, usually called "forams" or "bugs", comprise a large group. Some 60-80,000 species have been described from Cambrian through Recent age sediments. The species generally have shells made of silt and sand particles or secreted calcium carbonate that range in size from less than .1 mm to 10 cm. Most are about the size of a pin head. Each shell has one or more chambers in which the protoplasm of the living protist resided.

Forams live today from the shallowest intertidal zones to the deepest trenches of the oceans. A very few live in salty lakes and springs, where they have been transported by birds or other means. About 4,000 species are alive today. Of these, only 40 float in the water and are planktonic, the rest are benthic and live on the bottom of the ocean or on plants or animals. Some even live on other forams! Forams eat a wide variety of food, from bacteria through algae to various kinds of animals and other protists.

Forams are among the most abundant fossils. They first occur as simple tubes of sand in the Cambrian. Later more complex tubes and coils appear, even developing chambers in the Ordovician. In the Silurian, secreted calcium carbonate types appear; these diversify into many different shapes and kinds. Like animals and plants, they undergo extinctions at particular times in the geologic past, but radiate again into new, but similar shaped, forms. They remain abundant and diverse. Because of this, forams are the principal microfossil used to age-date and correlate marine sedimentary rocks — they are particularly useful in the oil industry. A single oil company, during the boom days of exploration for new deposits, might have employed 40 or 50 micropaleontologists to study forams. They are used also to decipher ancient environments, climates and oceanography. They are probably the most important fossils of any, simply because they are so useful.

Radiolaria

"Rads", as they are commonly known in paleontology, are marine protists with siliceous skeletons of rods and lattices arranged in many complex ways. Thousands of species of fossil forms have been described in the scientific literature, and hundreds are alive today. They first appear in the Cambrian as spherical lattices with a few rods sticking beyond the sphere. In the later Paleozoic, they also radiated into many species, and except for periods of extinction continued to modern times. They too have been useful in age-dating and correlation, but mostly of deep-sea sedimentary rocks. Rads have always been marine and all, as far as we know, are planktonic. They are especially abundant in regions of upwelling where phytoplankton that they feed on are common. Rads are generally smaller than forams and look glassy.

Diatoms

Diatoms are common marine and freshwater algae. They photosynthesize, hence only need certain nutrients and sunlight to survive and to make their siliceous skeletons. These fossilized skeletons, known as frustules, extend back only to the Jurassic or Cretaceous, although molecular biological evidence indicates that they arose in the Precambrian. Perhaps the missing record results from a lack of skeletons or a failure to preserve. In some places, thousands of feet of rock may be composed principally of diatom remains. These rocks, called diatomites, are usually light gray or white, fairly soft, and commonly well-bedded. Diatomites are mined for use as fine abrasives, filter materials for swimming pools and beer filtration, color carrier in cosmetics, toothpaste, and over a hundred other things.

Diatoms range in size from quite small to specks large enough to see with the naked eye. Their frustules are in two parts, one fitting inside the other like a petri dish. Diatoms live both in the water column and on a solid substrate — some even live on the skin of whales or between sand grains on beaches! The planktonic diatoms usually are circular or trigonal, while the benthic ones are elongated with a groove down one side. The groove or raphe allows protoplasm to extrude and move the entire diatom along the substrate. Diatoms, throughout their geologic history, have been most common in areas of shallow water or upwelling where nutrients are carried into the euphotic zone.

Most diatoms are generally easy to study. The methods of freeing them from rocks are the same in a classroom as for other microfossils. Once disaggregated from their host rock, they can be spread on a glass slide or dark piece of cardboard and viewed in transmitted or reflected light. They can be moved around with a very small, damp paintbrush. In the 1800's, people used to make designs and pictures on glass slides by placing different kinds of diatoms in particular places.

Animals

Nearly all sedimentary rocks have microfossils of animals and plants in them. Many times they are unidentifiable. Others can be readily attributed to a known group. For examples, fish scales, teeth, and bones are common in marine fine grained rocks, and small teeth and bones of terrestrial vertebrates can be found in nonmarine rocks. Most common, however, are small parts of invertebrates — shells, crinoid columnals, echinoid spines and plates, sponge spicules, ostracodes, and others. Some of these are not too useful in paleontology, but the tiny tooth-like structures known as conodonts are used for precise age-dating and correlation of rocks ranging in age from Cambrian to Triassic and ostracodes are useful in all respects from the Paleozoic to the Recent. Both conodonts and ostracodes make good classroom materials.

Plants

Bits and pieces of various plants are fairly common in sedimentary rocks. These are usually cuticle and other small parts. Seeds may be encountered in microfossil samples taken from rocks deposited in terrestrial environments. Pollen and spores are very common microfossils but they require difficult extraction techniques using dangerous acids and high powered microscopes to see them. These are best demonstrated with prepared slides purchased from supply.

Siliceous Microfossils

China contain abundant siliceous microfossils. Based on their high content of siliceous hard parts these rocks are, in fact, biosiliceous sediments. In more clay-rich lithologies, diagenetic dissolution processes, however, caused a removal of siliceous hard parts and led to "pure mudstones" and *silicified claystones*, being apparently devoid of siliceous microfossils. Abundant preservation as Fe-hydroxide pseudomorphs and in early diagenetic concretions do indicate, however, a high biosiliceous content of the original sediments. Rocks investigated belong to the lowermost Cambrian (*Anabarites trisulcatus – Protohertzina anabarica* assemblage zone) and come from black-chert sequences as well as (black shales (Nuititang Formation, Songtao Section, Eastern Guizhou Province; Niutitang Formation near the Mengdong train station, Hunan Province). Lighter coloured clay-rich lithologies of a similar age in Guizhou have similarly yielded abundant siliceous hard parts.

Siliceous microfossils are almost exclusively sponge spicules. Lithologically there seem to be strong similarities with coeval sedimentary rocks in southern Kazakhstan, termed *spongiolites*. Among the sponge spicules are megascleres as well as microscleres. Both spicule types allow clear identification of which particular sponge groups were present in the respective depositional environment. Remarkably, some of the solution residues consist exclusively of spicules derived from lithistid demosponges (especially in the lighter coloured, clay-rich facies), whereas others (*e.g.* most *black* shale and black chert samples) contain almost exclusively spicules of hexactinellid sponges. Attempting to relate sedimentary facies to the preservation of particuar sponge types, it appears that biofacial and ecologic differences were already present by the earliest Cambrian. Quite possibly, sponge spicule associations across the Precambrian-Cambrian boundary (with its predominantly siliceous and clay-rich lithologies on the Yangtze Platform) are suitable for a broadly applicable ecostratigraphic parallelization of rock units, for example in the Mesozoic of the Northern alps (Mostler, 1971). They also may eventually be used for biostratigraphic purposes across the Precambrian-Cambrian boundary.

Rocks rich in siliceous microfossils from the above mentioned occurrences represent sediments deposited on a marine shelf or slope (*transitional*). In the deeper oceanic basins represented on the Yangtze Platform (the depositional environment of the *Hetang Shales*, siliceous sponge remains are not common in solution residues. Eventually, the deeper water sea bottoms, being strongly anoxic had not yet been colonised by sponge faunas in the earliest Cambrian, or were only temporarily occupied during time periods of more favourable ecological conditions. These basinal lithologies, however, do contain strong petrographical as well as micropaleontological signs of extensive phytoplankton blooms occurring in the earliest Cambrian black shale depositional basins.

Somewhat astonishingly, remains of radiolarians are extremely rare in the solution residues. Only one deeper water sequence (the *Hetang* Shales in the Xintangwu Section, western Zhejiang Province, cf. Braun and Chen, 2003) yielded a few, very fragmentary specimens. These display strong similarities to spherical radiolarian morphologies found in Paleozoic rocks younger than Cambrian and Ordovician. However, further conclusions concerning early radiolarian evolution must await additional and more complete discoveries. Micropaleontological processing of chert successions in the Neoproterozoic and early.

Cambrian in oceanic basinal facies is expected to yield further data, as they are lithologically very similar to bedded radiolarian cherts of younger Paleozoic ages.

The significant contribution of biosiliceous particles to early Cambrian sedimentation on the Yangtze platform (as well as to other areas in Kazakhstan and Europe) implies that silica-biomineralizing organisms were playing a significant role in the geochemical cycling of silica in the oceans by the beginning of the Phanerozoic. The abundance of sponge spicules in the sediments indicate that this group (Porifera) played a major ecological role in the early Cambrian. This is supported by findings of complete sponges and spicule clusters on bedding planes of clay-rich sediments and black shales. But one must be cautious of any statement about the importance of siliceous plankton (radiolarians) by Early Cambrian times until further work is carried out.

Phosphatic microfossils and phosphatic particles from black shales and bedded black cherts contribute to the ecological picture of the oceanic basins as well as to the environments of the anoxic sea bottoms during earliest Cambrian times. The common occurrence of *Protohertzina* and microcoprolite-shaped phosphatic particles seems to support suggestions that the "less anoxic" near surface water masses above shelf and deeper-water areas, that contained rich phytoplankton could have served as a food source not only for a rich benthic sponge fauna, but also for metazoan plankton in the earliest Cambrian.

CHAPTER–4

Charophytes

INTRODUCTION

Charophytes or stoneworts are one of the largest and most structurally complex of the green algae, and are thought to be part of the evolutionary lineage that lead to vascular plants. As green algae, charophytes are sometimes classed within the Chlorophyta but are often classed as a distinct group, the Charophyta. These plants are aquatic, and modern forms are found in many freshwater to brackish habitats such as ponds, lakes, lagoons and streams. The main axes of mature plants comprise a series of multicellular nodes interspersed by relatively long single cells or internodes. Branches and branchlets grow from individual nodal cells in a whorled arrangement (see inset below). The gametangia, or fertile elements of the plant are quite characteristic in charophytes. The female gametangium or oogonium is formed from a number of divided cells that grow out from one of the nodes; the outer most cells are elongate and spiral around the oogonium forming a protective sheath (see inset below). The male gametangium or antheridium also grows from a nodal cell and is a complex spherical structure usually comprising eight shield-cells, each one enclosing a stalked cell bearing tiny rounded head-cells which in turn bear filaments comprising the cells which produce the spermatozoids. The position of the gametangia on the plants is variable with different charophyte species and can be diagnostic.

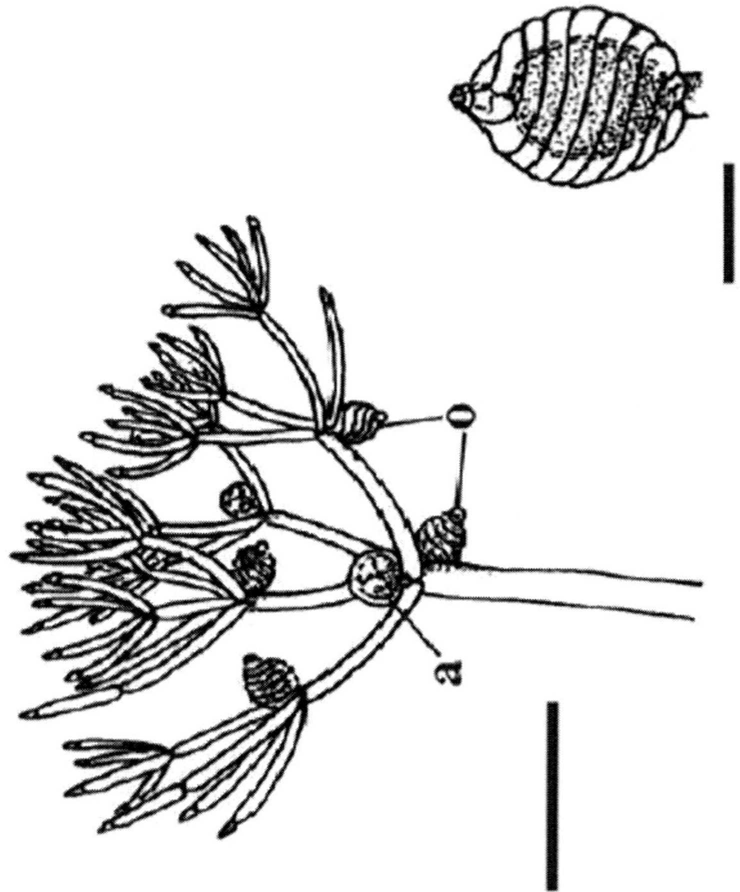

Fig.4.1: Above: Line drawings of a fertile axis of the extant *Nitella gracilis* showing oogonia (**o**) and antheridia (**a**) (left) (scale bar = 1mm) and close-up of an oogonium (right) (scale bar = 250μm) (based on Groves & Bullock-Webster 1920).

As fossils, charophytes can be quite in sediments deposited in suitable environments. It should be noted that the oogonia may be allochthonous or easily transported, but in instances where the axes of the plant are also preserved it is likely these are more-or-less *in situ* or autochthonous.

FOSSIL RECORD

Charophytes are quite well represented in the fossil record from the Tertiary, Cretaceous and Jurassic where they are locally abundant, particularly in limestones and marls deposited in brackish or freshwater settings. In these rocks it is usually the fossilised oogonia that are found (commonly called gyrogonites). This is because the oogonia are often calcified and are easily preserved as a mineral of calcium carbonate, typically calcite. In some instances, for example in the Upper Jurassic/Lower Cretaceous Purbeck Limestones of southern England, fossil charophyte oogonia are so common, diverse and widespread they can be used as zone fossils for biostratigraphy. The preservation of whole plants is not common, but they can be quite exquisite in their detail.

The earliest fossil charophytes have been recorded from the Upper Silurian, though whether or not these are true charophytes is still debated. Early Devonian charophytes have been found exquisitely preserved in the Rhynie chert and were first recorded. The Rhynie charophytes show many similarities with the extant charophyte group, the Nitelleae, and have been assigned to the species *Palaeonitella cranii*. The morphology of *Palaeonitella* is outlined below.

MORPHOLOGY

The morphology of *Palaeonitella* is relatively simple, its structure being closely comparable to that of modern Nitelleae (see inset above). An entire reconstruction of the alga has not been published, though much of the plant is known, primarily the branch whorls, rhizoid nodes and possible bulbils. The status of *Palaeonitella* has been questioned, but recent discoveries of fertile elements in the chert indicate unequivocally that the plant is a charophyte alga. The fertile elements will be figured here once published.

Axes

The overall size of the plant is not known, but it comprises upright primary axes that bear regularly spaced clusters of nodal cells with whorls of lateral branches. The spacing of nodal cells or rather the length of the internodes does, however, tend to decrease towards the distal end of the axes. These branches in turn may bear secondary branches. The number of branches or rays in a whorl can be variable. Whorls commonly bearing up to ten secondary rays have been observed on the lateral branches. The primary axes are slender generally being up to 200μm in diameter, lateral branches are conspicuously narrower being up to 100μm in diameter with very fine secondary branches being up to 50μm in diameter. Occasionally the nodal and internodal cells appear hypertrophic; mutated and swollen where they have been infested by aquatic fungi.

RHIZOIDS AND BULBILS

At the bases of the primary axes occur tubular cells occasionally displaying oblique septa and nodes. These structures represent the rhizoids of *Palaeonitella*. The nodal cells associated with the septa commonly bear small rhizoidal branches. These structures are closely comparable to those seen in the rhizoids of extant Nitelleae. Occasionally, conspicuous oval or spherical vesicular structures are seen in connection with these nodal cells. They vary in size, typically ranging between 200μm and 500μm in diameter and represent bulbils, another feature seen in some contemporary charophytes.

PALAEOECOLOGY

Palaeonitella was undoubtedly an aquatic plant. It is typically found in chert associated with other aquatic biota, particularly crustaceans, chlorophytic algae and chytrids (tiny aquatic fungi). The enclosing matrix commonly contains coprolites and exhibits a clotted or 'mulm-like' texture. It is likely that the alga was an early coloniser, primarily living in still, relatively shallow freshwater temporary ponds, apparently with rather soft silty and organic-rich substrates. If *Palaeonitella* had a similar ecology to that of most modern Nitelleae it is likely the waters it inhabited were generally quite alkaline. Most extant charophytes tolerate waters with a range of between pH 6 and pH 9. In a few charophyte-bearing chert horizons, many of the specimens preserved *in situ* exhibit an orientation with axes appearing to be swept or bent over in a particular direction. This probably represents current alignment (Fayers and Trewin in press); maybe *Palaeonitella* could also tolerate areas of slow moving water, perhaps colonising rather sheltered areas in the cool distal reaches of run-off channels from hot springs.

Chrysophytes, or golden algae, are common microscopic chromists in fresh water. Some species are colorless, but the vast majority are photosynthetic. As such, they are particularly important in lakes, where they may be the primary source of food for zooplankton. They are not considered truly autotrophic by some biologists because nearly all chrysophytes become facultatively heterotrophic in the absence of adequate light, or in the presence of plentiful dissolved food. When this occurs, the chrysoplast atrophies and the alga may turn predator, feeding on bacteria or diatoms.

There are more than a thousand described species of golden algae, most of them free-swimming and unicellular, but there are filamentous and colonial forms. Other chrysophytes may spend part of their life as amoeboid cells. At the left and center of the above illustration is *Dinobryon*, a freshwater genus in which the individual cells are surrounded by vase-shaped loricae, composed of chitin fibrils and other polysaccharides. The colonies grow as branched or unbranched chains. A spherical colonial form, *Synura*, is on the right; the surfaces of these cells are covered by silica scales. Species which produce siliceous coverings may have bristles or scales with quite complex structure. Some researchers group the chrysophytes with silica scales in a separate taxon, the Synurophyceae.

The oldest known chrysophytes are from calcareous and siliceous deposits of Cretaceous age, but they reached their greatest diversity in the Miocene. The group actually has a fairly complete fossil record, because most freshwater chrysomonads secrete resting cysts of silica, which may be abundant in certain rocks — in some Paleocene deposits, chrysophyte cysts outnumber the diatoms! The fossils of chrysophytes, like those of diatoms and coccolithophorids, are often used as paleoecological indicators to reconstruct ancient environments.

It is now generally believed that the Chrysophyta is a heterogeneous group, probably paraphyletic. Several groups formerly included here have been given separate recognition, such as the Raphidiophyceae, Eumastigophyceae, Xanthophyceae, Silicoflagellata, Sarcinochrysophyceae, and others. However, there is as yet no general consensus as to how these groups are related to each other or to the other chromist groups.

CHAPTER–5

Coccolith

INTRODUCTION

Coccoliths are individual plates of calcium carbonate formed by coccolithophores (single-celled algae such as *Emiliania huxleyi*) which are arranged around them in a *coccosphere.*

Formation and Composition

Coccoliths are formed within the cell in vesicles derived from the golgi body. When the coccolith is complete these vesicles fuse with the cell wall and the coccolith is excytosed and incorporated in the coccosphere. The coccoliths are either dispersed following death and breakup of the coccosphere, or are shed continually by some species. They sink through the water column to form an important part of the deep-sea sediments (depending on the water depth). Thomas Huxley was the first person to observe coccoliths in modern marine sediments and he gave them the name coccoliths. Coccoliths are composed of calcium carbonate as the mineral calcite and are the main constituent of chalk deposits such as the white cliffs of Dover.

TYPES

There are two main types of coccoliths, heterococcoliths and holococcoliths. Heterococcoliths are formed of a radial array of elaborately shaped crystal units. Holococcoliths are formed of minute (ca 0.1 micrometre) calcite rhombohedra arranged in continuous arrays. The two coccolith types were originally thought to be produced by different families of coccolithophores. Now, however, it is known through a mix of observations on field samples and laboratory cultures, that the two coccolith types are produced by the same species but at different life cycle phases. Heterococcoliths are produced in the diploid life-cycle phase and holococcoliths in the haploid phase.

FUNCTION

Although coccoliths are remarkably elaborate structures whose formation is a complex product of cellular processes, their function is unclear. Hypotheses include defence against grazing by zooplankton or infection by bacteria or viruses; maintenance of buoyancy; release of carbon dioxide for photosynthesis; to filter out harmful UV light; or in deep-dwelling species, to concentrate light for photosynthesis.

FOSSIL RECORD

Because coccoliths are formed of low-Mg calcite, the most stable form of calcium carbonate, they are readily fossilised. We find them in the sediment together with similar microfossils of uncertain affinities (nannoliths) from the Triassic to recent. Coccoliths and related fossils are referred to as *calcareous nannofossils* or *calcareous nannoplankton.*

CALCIUM CARBONATE

Calcium carbonate is a chemical compound with the chemical formula $CaCO_3$. It is a common substance found as rock in all parts of the world, and is the main component of shells of marine organisms, snails, and eggshells. Calcium carbonate is the active ingredient in agricultural lime, and is usually the principal cause of hard water. It is commonly used medicinally as a calcium supplement or as an antacid, but high consumption can be hazardous.

Occurrence

Calcium carbonate is found naturally as the following minerals and rocks:

- Aragonite
- Calcite

- Vaterite or (μ-$CaCO_3$)
- Chalk (Blackboard chalk is calcium sulfate, $CaSO_4$)
- Limestone
- Marble
- Travertine

To test whether a mineral or rock contains carbonate, strong acids, such as hydrochloric acid or sulfuric acid, can be added to it. (Although sulfuric acid reacts, the action soon ceases because the calcium sulfate produced is rather insoluble in water and limits the reaction.) If the sample does contain carbonate, it will fizz and produce carbon dioxide and water. Weak acids such as acetic acid will react, albeit less vigorously. All of the rocks/ minerals mentioned above will react with acid. To test for calcium, prepare a platinum or nichrome wire and dip it into some hydrochloric acid. Then, dip the wire into some crushed sample to be tested. Place the wire in a Bunsen Flame, if calcium is presented in the sample, brick-red flame will be produced. If a sample gives positive results for both of the two tests above, it is calcium carbonate.

CHEMICAL PROPERTIES

Calcium carbonate shares the typical properties of other carbonates. Notably:

it reacts with strong acids, releasing carbon dioxide:

$$CaCO_{3(s)} + 2HCl_{(aq)} \rightarrow CaCl_{2(aq)} + CO_{2(g)} + H_2O_{(l)}$$

it releases carbon dioxide on heating (to above 840°C in the case of $CaCO_3$), to form calcium oxide, commonly called quicklime, with reaction enthalpy 178 kJ/mole:

$$CaCO_3 \rightarrow CaO + CO_2$$

Calcium carbonate will react with water that is saturated with carbon dioxide to form the soluble calcium bicarbonate.

$$CaCO_3 + CO_2 + H_2O \rightarrow Ca(HCO_3)_2$$

This reaction is important in the erosion of carbonate rocks, forming caverns, and leads to hard water in many regions.

Preparation

The vast majority of calcium carbonate used in industry is extracted by mining or quarrying. Pure calcium carbonate (e.g. for food or pharmaceutical use), can be produced from a pure quarried source (usually marble).

Alternatively, calcium oxide is prepared by calcining crude calcium carbonate. Water is added to give calcium hydroxide, and carbon dioxide is passed through this solution to precipitate the desired calcium carbonate, referred to in the industry as precipitated calcium carbonate (PCC)

$$CaCO_3 \rightarrow CaO + CO_2$$

$$CaO + H_2O \rightarrow Ca(OH)_2$$

$$Ca(OH)_2 + CO_2 \rightarrow CaCO_3 + H_2O$$

Industrial Applications

The main use of calcium carbonate is in the construction industry, either as a building material in its own right (e.g. marble) or limestone aggregate for roadbuilding or as an ingredient of cement or as the starting material for the preparation of builder's lime by burning in a kiln.

Calcium carbonate is also used in the purification of iron from iron ore in a blast furnace. Calcium carbonate is calcined *in situ* to give calcium oxide, which forms a slag with various impurities present, and separates from the purified iron.

Calcium carbonate is also used in the oil industry in drilling fluids as a formation bridging and filtercake sealing agent and may also be used as a weighting material to increase the density of drilling fluids to control downhole pressures.

Calcium carbonate is also one of the main sources used in growing Seacrete, or Biorock.

Calcium carbonate is widely used as an extender in paints, in particular matte emulsion paint where typically 30% by weight of the paint is either chalk or marble.

Calcium carbonate is also widely used as a filler in plastics Some typical examples include around 15 to 20% loading of chalk in uPVC drain pipe, 5 to 15% loading of stearate coated chalk or marble in uPVC window profile. PVC cables can use calcium carbonate at loadings of up to 70 phr (parts per hundred parts of resin) to improve mechanical properties (tensile strength and elongation) and electrical properties (volume resistivity). Polypropylene compounds are often filled with calcium carbonate to increase rigidity, a requirement that becomes important at high use temperatures. It also routinely used as a filler in thermosetting resins (Sheet and Bulk moulding compounds) and has also been mixed with ABS, and other ingredients, to form some types of compression molded "clay" Poker chips.

Fine ground calcium carbonate is an essential ingredient in the microporous film used in babies' diapers and some building films as the pores are nucleated around the calcium carbonate particles during the manufacture of the film by biaxial stretching.

Calcium carbonate is also used in a wide range of trade and DIY adhesives, sealants, and decorating fillers Ceramic tile adhesives typically contain 70 to 80% limestone. Decorating crack fillers contain similar levels of marble or dolomite. It is also mixed with putty in setting stained glass windows, and as a resist to prevent glass from sticking to kiln shelves when firing glazes and paints at high temperature.

Calcium carbonate is known as *whiting* in ceramics/glazing applications, where it is used as a common ingredient for many glazes in its white powdered form. When a glaze containing this material is fired in a kiln, the whiting acts as a flux material in the glaze.

In North America, calcium carbonate has begun to replace kaolin in the production of glossy paper. Europe has been practicing this as alkaline papermaking or acid-free papermaking for some decades. Carbonates are available in forms: ground calcium carbonate (GCC) or precipitated calcium carbonate (PCC). The latter has a very fine and controlled particle size, on the order of 2 micrometres in diameter, useful in coatings for paper.

Used in swimming pools as a pH corrector for maintaining alkalinity "buffer" to offset the acidic properties of the disinfectant agent.

It is commonly called chalk as it has been a major component of blackboard chalk. Chalk may consist of either calcium carbonate or gypsum, hydrated calcium sulfate $CaSO_4 \cdot 2H_2O$.

Health and Dietary Applications

Calcium carbonate is widely used medicinally as an inexpensive dietary calcium supplement or antacid It may be used as a phosphate binder for the treatment of hyperphosphatemia (primarily in patients with chronic renal failure). It is also used in the pharmaceutical industry as an inert filler for tablets and other pharmaceuticals. Calcium carbonate is also used in homeopathy as one of the constitutional remedies.

Excess calcium from supplements, fortified food and high-calcium diets, can cause the "milk alkali syndrome," which has serious toxicity and can be fatal. In 1915, Bertram Sippy introduced the "Sippy regimen" of hourly ingestion of milk and cream, and the gradual addition of eggs and cooked cereal, for 10 days, combined with alkaline powders, which provided symptomatic relief for peptic ulcer disease. Over the next several decades, the Sippy regimen resulted in renal failure, alkalosis, and hypercalemia, mostly in men with peptic ulcer disease. These adverse effects were reversed when the regimen stopped, but it was fatal in some patients with protracted vomiting. Milk alkali syndrome declined in men after effective treatments for peptic ulcer disease. But during the past 15 years, it has been reported in women taking calcium supplements above the recommended range of 1200 to 1500 mg daily, for prevention and treatment of osteoporosis, and is exacerbated by dehydration. Calcium has been added to over-the-counter products, which contributes to inadvertent excessive intake. Excessive calcium intake can lead to hypercalcemia, complications of which include vomiting, abdominal pain and altered mental status.

A form of food additive is designated as E170. It is used in some soy milk products as a source of dietary calcium; one study suggests that calcium carbonate might be bioavailable as the calcium in cow's milk.

Ecological Applications

In 1989, a researcher, Ken Simmons, introduced $CaCO_3$ into the Whetstone Brook in Massachusetts. His hope was that the calcium carbonate would counter the acid in the stream from acid rain and save the trout that had ceased to spawn. Although his experiment was a success, it did increase the amounts of aluminium ions in the area of the brook that was not treated with the limestone. This shows that $CaCO_3$ can be added to neutralize the effects of acid rain in river ecosystems. Currently calcium carbonate is used to neutralize acidic conditions in both soil and water.

Calcination Equilibrium

Calcination of limestone using charcoal fires to produce quicklime has been practiced since antiquity by cultures all over the world. The temperature at which limestone yields calcium oxide is usually given as 825 °C, but stating an absolute threshold is misleading. Calcium carbonate exists in equilibrium with calcium oxide and carbon dioxide at any temperature. At each temperature there is a partial pressure of carbon dioxide that is in equilibrium with calcium carbonate. At room temperature the equilibrium overwhelmingly favors calcium carbonate, because the equilibrium CO_2 pressure is only a tiny fraction of the partial CO_2 pressure in air, which is about 0.035 kPa.

At temperatures above 550°C the equilibrium CO_2 pressure begins to exceed the CO_2 pressure in air. So above 550°C, calcium carbonate begins to outgas CO_2 into air. But in a charcoal fired kiln, the concentration of CO_2 will be much higher than it is in air. Indeed if all the oxygen in the kiln is consumed in the fire, then the partial pressure of CO_2 in the kiln can be as high as 20 kPa.

For the outgassing of CO_2 from calcium carbonate to happen at an economically useful rate, the equilibrium pressure must significantly exceed the ambient pressure of CO_2. And for it to happen rapidly, the equilibrium pressure must exceed total atmospheric pressure of 101 kPa, which happens at 898°C.

CLAW HYPOTHESIS

The CLAW hypothesis proposes a feedback loop that operates between ocean ecosystems and the Earth's climate. The hypothesis specifically proposes that particular phytoplankton that produce dimethyl sulfide are responsive to variations in climate forcing, and that these responses lead to a negative feedback loop that acts to stabilise the temperature of the Earth's atmosphere. The CLAW hypothesis was originally proposed by Robert Charlson, James Lovelock, Meinrat Andreae and Stephen Warren, and takes its acronym from the first letter of their surnames.

The hypothesis describes a feedback loop that begins with an increase in the available energy from the sun acting to increase the growth rates of phytoplankton by either a physiological effect (due to elevated temperature) or enhanced photosynthesis (due to increased irradiance). Certain phytoplankton, such as coccolithophorids, synthesise dimethylsulfoniopropionate (DMSP), and their enhanced growth increases the production of this osmolyte. In turn, this leads to an increase in the concentration of its breakdown product, dimethyl sulfide (DMS), in first seawater, and then the atmosphere. DMS is oxidised in the atmosphere to form sulfur dioxide, and this leads to the production of sulfate aerosols. These aerosols act as cloud condensation nuclei and increase cloud droplet number, which in turn elevate the liquid water content of clouds and cloud area. This acts to increase cloud albedo, leading to greater reflection of incident sunlight, and a decrease in the forcing that initiated this chain of events. The figure to the right shows a summarising schematic diagram. Note that the feedback loop can operate in reverse, such that a decline in solar energy leads to reduced cloud cover and thus to an increase in the amount of solar energy reaching the Earth's surface.

A signifcant feature of the chain of interactions described above is tht it creates a negative feedback loop, whereby a change to the clmate system (increased/decreased solar input) is ultimately counteracted and damped by the loop. As such, the CLAW hypothesis posits an example of planetary-scale homeostasis or complex adaptive system, consistent with the Gaia hypothesis framed by one of the original authors of the CLAW hypothesis, James Lovelock.

Some subsequent studies of the CLAW hypothesis have uncovered some evidence to support its mechanism although this is not unequivocal Other researchers have suggested that a CLAW-like mechanism may operate in the Earth's sulfur cycle without the requirement of an active biological component.

THE ANTI-CLAW HYPOTHESIS

In a recent book, *The Revenge of Gaia,* Lovelock has proposed that instead of providing negative feedback in the climate system, the components of the CLAW hypothesis may act to create a positive feedback loop.

Under future global warming, increasing temperature may stratify the world ocean, decreasing the supply of nutrients from the deep ocean to its productive euphotic zone. Consequently, phytoplankton activity will decline with a concommitant fall in the production of DMS. In a reverse of the CLAW hypothesis, this decline in DMS production will lead to a decrease in cloud condensation nuclei and a fall in cloud albedo. The consequence of this will be further climate warming which may lead to even less DMS production (and further climate warming...)

Evidence for the anti-CLAW hypothesis is constrained by similar uncertainties as those of the sulfur cycle feedback loop of the CLAW hypothesis. However, researchers simulating future oceanic primary production have found evidence of declining production with increasing ocean stratification.

CHAPTER–6

Conodont and Chordate

CONODONT

Introduction

Conodonts are extinct chordates resembling eels, classified in the class Conodonta. For many years, they were known only from tooth-like microfossils now called conodont elements, found in isolation. The animal is also called conodontophora (conodont bearers) to avoid ambiguity.

Description

The eleven known fossil imprints of conodont animals depict an eel-like creature with 15 or, more rarely, 19 elements forming a bilaterally symmetrical array in the head. This array comprised a feeding apparatus radically different from the jaws of modern animals. There are three forms of teeth, *coniform* cones, *ramiform* bars, and *pectiniform* platforms, which may have performed different roles.

The organisms range from a centimeter or so to the giant *Promissum*, 40cm in length It is now widely agreed that conodonts had large eyes, fins with fin rays, chevron-shaped muscles and a notochord.

Ecology

The "teeth" of some conodonts have been interpreted as filter-feeding apparatuses, filtering out plankton from the water and passing it down the throat. Others have been interpreted as a "grasping and crushing array".

The lateral position of the eyes makes a predatory role unlikely.

The preserved musculature hints that some conodonts (*Promissum* at least) were efficient cruisers but incapable of bursts of speed.

Classification

The conodonts are currently classified in the phylum Chordata because their fins with fin rays, chevron-shaped muscles and notochord are characteristic of Chordata.

They are considered by Milsom and Rigby to be vertebrates similar in appearance to modern hagfish and lampreys and phylogenetic analysis suggests that they are more derived than either of these groups. This analysis, however, comes with one caveat: early forms of conodonts, the protoconodonts, appear to form a distinct clade from the later paraconodonts and euconodonts. It appears likely that the protoconodonts represent a stem group to the phylum containing chaetognath worms, indicating that they are not close relatives of true conodonts.

Conodont Teeth Fossils

For many years, conodonts were known only from enigmatic tooth-like microfossils, which occur commonly but not always in isolation, and were not associated with any other fossil. These phosphatic microfossils are now termed "conodont elements" to avoid confusion. This confusion is most apparent for the non-specialist in the book "Your Inner Fish", by Neil Shubin, who describes the origin of teeth in chapter 4. In this chapter, the author attaches the name "conodont" to both the "conodont bearer" (the animal) and the "conodont elements" (the teeth), and the reader may have a hard time to make sense of the concept of "animals living in the mouths of animals".

They are widely used in biostratigraphy.

Conodont elements are also used as paleothermometers, a proxy for thermal alteration in the host rock. This is because under higher temperatures the phosphate undergoes predictable and permanent color changes, measured with the conodont alteration index. This has made them useful for petroleum exploration where they are known, in rocks dating from the Cambrian to the Late Triassic.

It was not until early 1980s that the conodont teeth were found in association with fossils of the host organism, in a konservat lagerstätte. This is because most of the conodont animal was soft-bodied, thus everything but the teeth were not suited for preservation under normal circumstances.

CHORDATE

Chordates (phylum Chordata) are a group of animals that includes the vertebrates, together with several closely related invertebrates. They are united by having, at some time in their life cycle, a notochord, a hollow dorsal nerve cord, pharyngeal slits, an endostyle, and a post-anal tail. The phylum Chordata consists of three subphyla: Urochordata, represented by tunicates; Cephalochordata, represented by lancelets;and Craniata, which includes Vertebrata. The Hemichordata have been presented as a fourth chordate subphylum, but they are now usually treated as a separate phylum. Urochordate larvae have a notochord and a nerve cord but these are lost in adulthood. Cephalochordates have a notochord and a nerve cord but no brain or specialist sense organs, and a very simple circulatory system. Craniates are the only sub-phylum whose members have skulls. In all craniates except for Hagfish, the dorsal hollow nerve cord has been surrounded with cartilaginous or bony vertebrae and the notochord generally reduced; hence hagfish are not regarded as vertebrates. The chordates and three sister phyla, the Hemichordata, the Echinodermata and the Xenoturbellida, make up the deuterostomes, one of the two superphyla which encompass all fairly complex animals.

Attempts to work out the evolutionary relationships of the chordates have produced several hypotheses, but the current consensus is that chordates are monophyletic, in other words contain all and only the descendants of a single common ancestor *which is itself a chordate,* and that craniates' nearest relatives are cephalochordates. All of the earliest chordate fossils have been found in the Early Cambrian Chengjiang fauna, and include two species that are regarded as fish, which implies that these are vertebrates. Because the fossil record of chordates is poor, only molecular phylogenetics offers a reasonable prospect of dating their emergence. However the use of molecular phylogeneticss for dating evolutionary transitions is controversial.

It has also proved difficult to produce a detailed classification within the living chordates. Attempts to produce evolutionary "family trees" give results that differ from traditional classes because several of those classes are not monophyletic. As a result vertebrate classification is in a state of flux.

Definition

Chordates form a phylum - a grouping of animals with a shared bodyplan - defined by having at some stage in their lives all of the following:

- a notochord, in other words a fairly stiff rod of cartilage that extends along the inside of the body and helps the animal to swim by flexing its tail.
- a dorsal neural tube, which develops into the spinal cord, the main commmunications trunk of the nervous system, in fish and other vertebrates.
- pharyngeal slits. The pharynx is the part of the throat immediately behind the mouth. In fish the slits are modified to form gills, but in other chordates they are part of a filter feeding system that extracts particles of food from the water in which the animals live.
- a muscular tail that extends backwards behind the anus.
- an endostyle. This groove in the ventral wall of the pharynx produces mucus to gather food particles, helps in transporting food to the esophagus, and stores iodine. It may be a precursor of the vertebrate thyroid gland.

Sub-divisions

There are three major groupings within the chordates:

Craniates have distinct skulls. Michael J. Benton comments that "craniates are characterized by their heads, just as chordates, or possibly all deuterostomes, are by their tails." Most are vertebrates, in which the notochord is relaced by the spinal column. This consists of a series of bony or cartilaginous cylindrical vertebrae, generally with neural arches that protect the spinal chord and with projections that link the vertebrae. Hagfish have incomplete braincases and no vertebrae, and are therefore not regarded as vertebrates but as members of the craniates, the group from which vertebrates are thought to have evolved. The position of lampreys is ambiguous. They have complete braincases and rudimentary vertebrae, and therefore may be regarded as vertebrates and true fish However molecular phylogenetics, which uses biochemical features to classify organisms, has produced both results that group them with vertebrates and others that group them with hagfish.

Cephalochordates are small, "vaguely fish-shaped" animals that lack brains, clearly-defined heads and specialized sense organs. These burrowing filter-feeders may be either the closest living relatives of craniates or surviving members of the group from which all other chordates evolved.

Most tunicates appear as adults in two major forms, both of which are bags of jelly that lack the standard features of chordates: "sea squirts" are sessile and consist mainly of water pumps and filter feeding apparatus; salps float in mid-water, feeding on plankton, and have a two-generation cycle in which one generation is solitary and the next forms chain-like colonies However all tunicate larvae have the standard chordate features, including long, tadpole-like tails; they also have rudimentary brains, light sensors and tilt sensors. The third main group of tunicates, Appendicularia (also known as Larvacea) retain tadpole-like shapes and active swimming all their lives, and were for a long time regarded as larvae of sea squirts or salps Because of their larvae's long tails tunicates are also called urochordates ("tail chordates").

Closest Non-chordate Relatives

Hemichordates ("half chordates") have some features similar to those of chordates: branchial openings that open into the pharynx and look rather like gill slits; stomochords, similar in composition to notochords but running in a circle round the "collar", which is ahead of the mouth; and a dorsal nerve cord – but also a smaller ventral nerve cord. There are two living groups of hemichordates. The solitary enteropneusts, commonly known as "acorn worms", have long probosces and worm-like bodies with up to 200 branchial slits, are up to 2.5 metres (8.2 ft) long, and burrow though seafloor sediments. Pterobranchs are colonial animals, often less than 1 millimetre (0.039 in) long individually, whose dwellings are inter-connected. Each filter feeds by means of a pair of branched tentacles, and has a short, shield-shaped proboscis. The extinct graptolites, colonial animals whose fossils look tiny hacksaw blades, lived in tubes similar to those of pterobranchs.

Echinoderms differ from chordates' other relatives in three conspicuous ways: instead of having bilateral symmetry they have radial symmetry, like wheels; their bodies are supported by skeletons made of calcite, a material not used by chordates, and these skeleton enclose their bodies but are also covered by a thin skin; they have tube feet. The feet are powered by another unique feature of echinoderms, a water vascular system of canals that also function as a "lung" and are surrounded by muscles that act as pumps. Crinoids look rather like flowers, and use their feather-like arms to filter food particles out of the water; most live anchored to rocks, but a few can move very slowly. Other echinoderms are mobile and take a variety of body shapes, for example starfish, sea urchins and sea cucumbers.

Origins

The majority of animals more complex than jellyfish and other Cnidarians are split into two groups, the protostomes and deuterostomes, and chordates are deuterostomes. It seems very likely that 555 million years old *Kimberella* was a member of the protostomes If so, this means that the protostome and deuterostome lineages must have split some time before

Kimberella appeared - at least 558 million years ago, and hence well before the start of the Cambrian 542 million years ago. The Ediacaran fossil *Ernettia*, from about 549 to 543 million years ago, may represent a deuterostome animal.

Fossils of one major deuterostome group, the echinoderms (whose modern members include starfish, sea urchins and crinoids) are quite common from the start of the Cambrian, 542 million years ago The Mid Cambrian fossil *Rhabdotubus johanssoni* has been interpreted as a pterobranch hemichordate. Opinions differ about whether the Chengjiang fauna fossil *Yunnanozoon*, from the earlier Cambrian, was a hemichordate or chordate. Another Chenjiang fossil, *Haikouella lanceolata*, also from the Chengjiang fauna, is interpreted as a chordate and possibly a craniate, as it shows signs of a heart, arteries, gill filaments, a tail, a neural chord with a brain at the front end, and possibly eyes - although it also had short tentacles round its mouth *Haikouichthys* and *Myllokunmingia*, also from the Chenjiang fauna, are regarded as fish. *Pikaia*, discovered much earlier but from the Mid Cambrian Burgess Shale, is also regarded as a primitive chordate. On the other hand fossils of early chordates are very rare, since non-vertebrate chordates have no bones or teeth, and none have been reported for the rest of the Cambrian.

The evolutionary relationships between the chordate groups and between chordates as a whole and their closest deuterostome relatives have been debated since 1890. Studies based on anatomical, embryological, and paleontological data have produced different "family trees". Some closely linked chordates and hemichordates, but that idea is now rejected Combining such analyses with data from a small set of ribosome RNA genes eliminated some older ideas, but open the possibility that tunicates (urochordates) are "basal deuterostomes", in other words surviving members of the group from which echinoderms, hemichordates and chordates evolved. Most researchers agree that, within the chordates, craniates are most closely related to cephalochordates, but there also reasons for regarding tunicates (urochordates) as craniates' closest relatives one other phylum,

Xenoturbellida, appears to be basal within the deuterostomes, in other words closer to the original deuterostomes than to the chordates, echinoderms and hemichordates.

Since chordates have left a poor fossil record, attempts have been made to calculate the key dates in their evolution by molecular phylogenetics techniques, in other words by analysing biochemical differences, mainly in RNA. One such study suggested that deuterostomes arose before 900 million years ago and the earliest chordates around 896 million years ago. However molecular estimates of dates often disagree with each other and with the fossil record, and their assumption that the molecular clock runs at a known constant rate has been challenged.

Classification

Taxonomy

The following schema is from the third edition of *Vertebrate Palaeontology* While it is structured so as to reflect evolutionary relationships (similar to a cladogram), it also retains the traditional ranks used in Linnaean taxonomy.

- **Phylum Chordata**
- — Subphylum Tunicata (Urochordata)— (tunicates, 3,000 species)
- — Subphylum Cephalochordata (Acraniata)— (lancelets, 30 species)
- — Subphylum Vertebrata (Craniata) (vertebrates — animals with backbones; 57,674 species)
  - – Class 'Agnatha' Paraphyletic (jawless vertebrates; 100+ species)
  - – Subclass Myxinoidea (hagfish; 65 species)
  - – Subclass Petromyzontida (Lampreys)
  - – Subclass Conodonta
  - – Subclass Pteraspidomorphi (Paleozoic jawless fish)
  - – Order Anaspida

- Order Thelodonti (Paleozoic jawless fish)

— Infraphylum Gnathostomata (jawed vertebrates)

- Class Placodermi (Paleozoic armoured forms)
- Class Chondrichthyes (cartilaginous fish; 900+ species)
  - Class Acanthodii (Paleozoic "spiny sharks")
  - Class Osteichthyes (bony fishes; 30,000+ species)
  - Subclass Actinopterygii (ray-finned fish; about 30,000 species)
  - Subclass Sarcopterygii (lobe-finned fish)
  - Superclass Tetrapoda (four-legged vertebrates; 18,000+ species)
  - Class Amphibia (amphibians; 6,000 species)
  - Series Amniota (with amniotic egg)
  - Class Sauropsida — (reptiles; 8,225+ species)
  - Class Aves (birds; 8,800–10,000 species)
  - Class Synapsida (mammal-like "reptiles"; 4,500 + species)
  - Class Mammalia (mammals; 5,800 species)

CHAPTER–7

Eel and Annelid

True eels (*Anguilliformes*) are an order of fish, which consists of four suborders, 19 families, 110 genera and approximately 600 species. Most eels are predators.

The flat and transparent larva of the eel is called a leptocephalus. A young eel is called an elver.

Description

True eels are elongated fishes, ranging in length from 5 centimetres (2.0 in) in the one-jawed eel (*Monognathus ahlstromi*) to 3.75 metres (12.3 ft) in the giant moray. They possess no pelvic fins, and many species also lack pectoral fins. The dorsal and anal fins are fused with the caudal or tail fin, to form a single ribbon running along much of the length of the animal.

Most eels prefer to dwell in shallow waters or hide at the bottom layer of the ocean, sometimes in holes. These holes are called eel pits. Only the Anguillidae family comes to fresh water to dwell there (not to breed). Some eels dwell in deep water (in case of family Synaphobranchidae, this comes to a depth of 4,000 metres (13,000 ft), or are active swimmers (the family Nemichthyidae - to the depth of 500 metres (1,600 ft).

Classification

This classification follows FishBase in dividing the eels into fifteen families. Additional families that are included in other classifications (notably ITIS and Systema Naturae 2000) are noted below the family with which they are synomized in the FishBase system.

Suborders and Families

Suborder: Anguilloidei

- Anguillidae (freshwater eels)
- Chlopsidae (false morays)
- Heterenchelyidae
- Moringuidae (spaghetti eels)
- Muraenidae (moray eels)
- Myrocongridae

Suborder: Congroidei

- Colocongridae
- Congridae (congers)
- Including Macrocephenchelyidae
- Derichthyidae (longneck eels)
- Including Nessorhamphidae
- Muraenesocidae (conger pikes)
- Nettastomatidae (witch eels)
- Ophichthidae (snake eels)

Suborder: Nemichthyoidei

- Nemichthyidae (snipe eels)
- Serrivomeridae (sawtooth eels)

Suborder: Synaphobranchoidei

- Synaphobranchidae (cutthroat eels)
- Including Dysommidae, Nettodaridae, and Simenchelyidae

In some classifications the family Cyematidae of bobtail snipe eels is included in the Anguilliformes, but in the FishBase system that family is included in the order Saccopharyngiformes.

The so-called "electric eel" of South America is not a true eel, but is more closely related to the Carp.

Use by Humans

Freshwater eels (*unagi*) and marine eels (conger eel, *anago*) are commonly used in Japanese cuisine - foods such as Unadon and Unajuu are popular but expensive. Eels are also very popular as food in Chinese cuisine, particularly Cantonese and Shanghai cuisine. Eel prices in Hong Kong often reached ¥1000 per kilogram, and even exceeded ¥5000 per kilogram at one time. Eel is also popular in Korean cuisine and is seen as a source of "stamina" for men. The European eel and other freshwater eels are eaten in Europe, the United States, and other places around the world. A traditional East London food is jellied eels. The Basque delicacy *angulas* consists of deep-fried elver (young eels). New Zealand longfin eel is a traditional food for Maori in New Zealand. In Italian cuisine eels from the Comacchio area (a swampy zone along the Adriatic coast) are specially prized along with the freshwater ones of the Bolsena Lake. In northern Germany and in The Netherlands, smoked eel is prized as a delicacy.

Eels are popular among marine aquarists in the United States, particularly the Moray eel which is commonly kept in tropical saltwater aquariums.

Fishing for eels is best done at night with either a gaff or a strong line with a small and square bloody piece of red meat.

Elvers were once eaten by fishermen as a cheap dish, but environmental changes have led to increased rarity of the fish. They are now considered a delicacy and are priced at up to £700 per kg in the United Kingdom.

Further Information

An urban legend states that wallets made out of electric eels (which, despite their name, are not eels) can demagnetize credit cards. This was proven to be untrue in an episode of the *MythBusters* TV show. As pointed out in the Straight Dope, eel-skin wallets are made from hagfish which are unrelated to electric eels. Furthermore, it seems that magnetic clasps, not eel leather, are to blame for demagnetization.

Eel blood is toxic, but the toxic protein it contains is destroyed by cooking. The toxin derived from eel blood serum was used by Charles Robert Richet in his Nobel winning research which discovered anaphylaxis (by injecting it into dogs and observing the effect).

On January 31, 1930, the Danish research ship "The Dana" captured what researchers believed to be a six-feet long eel larva near South Africa's Cape of Good Hope. This would have meant there were very long eels in the sea, since the typical eel larva is three inches (76 mm) long, while the adults can grow from about 4 feet (1.2 m) to 16 feet (4.9 m) long. In 1970, Dr. David G. Smith of the University of Miami identified the larva found as that of the spiny eel, an eel-like fish whose larval length is equal to its adult length, while the larval length of the true eel is much shorter than its adult length.

One of the famous attractions of the Pacific island of Huahine (part of the Society Islands in French Polynesia), is the bridge that crosses over a stream with 3- to 6-foot (1.8 m) long eels. These eels are deemed sacred by the locals, by local mythology.

Threats to Eels

There are strong concerns that the European eel population might be devastated by a new threat: *Anguillicola crassus*, a foreign parasitic nematode. This parasite from East Asia (the original host is *Anguilla japonica*) appeared in European eel populations in the early 1980s. Since 1995 it also appeared in the United States (Texas and South Carolina), most likely due to uncontrolled aquaculture eel shipments. In Europe, eel populations are already from 30% to 100% infected with the nematode. Recently it was shown that this parasite inhibits the function of the swimbladder as a hydrostatic organ (Wuertz et al. 1996). As an open ocean voyager, eels need the carrying capacity of the swimbladder (which makes up 3-6% of the eel's bodyweight) to cross the ocean on stored energy alone.

Because the eels are catadromous (living in fresh water but spawning in the sea), dams and other river obstructions can block their ability to reach inland feeding grounds. Since the 1970s an increasing number of eel ladders have been constructed in North America and Europe to help the fish bypass obstructions.

In New Jersey, an ongoing project monitors the glasseel migration with an online *in situ* microscope. As soon as more funding becomes available, it will be possible to log into the system via a Longterm Ecological Observatory (LEO) site.

ANNELID

The annelids, collectively called Annelida (from Latin *anellus* "little ring"), are a large phylum of animals comprising the segmented worms, with about 15,000 modern species including the well-known earthworms and leeches. They are found in most wet environments, and include many terrestrial, freshwater, and especially marine species (such as the polychaetes), as well as some which are parasitic or mutualistic. They range in length from under a millimeter to over three metres (the seep tube worm *Lamellibrachia luymesi*).

Physiology

Annelids are bilaterally symmetric and triploblastic protostomes with a coelom (which makes them coelomates), closed circulatory system and true segmentation. Their segmented bodies and coelom have given them evolutionary advantages over other worms. Oligochaetes and polychaetes typically have spacious coeloms; in leeches, the coelom is filled in with tissue and reduced to a system of narrow canals; archiannelids may lack the coelom entirely. The coelom is divided into a sequence of compartments by walls called septa. In the most general forms each compartment corresponds to a triple segment of the body, which also includes a portion of the nervous and (closed) circulatory systems, allowing it to function relatively independently. The closed circulatory system consists of networks of vessels containing blood with oxygen-carrying hemoglobin. Dorsal and ventral vessels are connected by segmental pairs of vessels. The dorsal vessel and five pairs of vessels that circle the esophagus of an earthworm are muscular and pump blood through the circulatory system. Tiny blood vessels are abundant in the earthworm's skin, which function as its respiratory organ. Each segment (metamere) is marked externally by one or more rings, called annuli. Each segment also has an outer layer of circular muscle underneath a thin cuticle and epidermis, and a system of longitudinal muscles. In earthworms and in daria the longitudinal muscles are strengthened by collagenous lamellae; the leeches have a double layer of muscles between the outer circulars and inner longitudinals. In most forms they also carry a varying number of bristles, called setae, and among the polychaetes a pair of appendages, called parapodia.

Anterior to the true segments lies the prostomium and peristomium, which carries the mouth, and posterior to them lies the pygidium, where the anus is located. The digestive tract is quite variable but is usually specialized. For example, in some groups (notably most earthworms) it has a typhlosole (to increase surface area) along much of its length. Different species of annelids have a wide variety of diets, including active and passive hunters, scavengers, filter feeders, direct deposit feeders which simply ingest the sediments, and blood-suckers. Annelids can also grow up to six inches.

The vascular system and the nervous system are separate from the digestive tract. The vascular system includes a dorsal vessel conveying the blood toward the front of the worm, and a ventral longitudinal vessel which conveys the blood in the opposite direction. The two systems are connected by a vascular sinus and by lateral vessels of various kinds, including in the true earthworms, capillaries on the body wall.

The nervous system has a nerve cord from which lateral nerves come in contact with each segment. Every segment has an autonomy; however, they unite to perform as a single body for functions such as locomotion. Growth in many groups occurs by replication of individual segmental units, in others the number of segments is fixed in early development.

Depending upon the species, annelids can reproduce both sexually and asexually.

Asexual Reproduction

Asexual reproduction by fission is a method used by some annelids and allows them to reproduce quickly. The posterior part of the body breaks off and forms a new identical worm. The position of the break is usually determined by an epidermal growth. *Lumbriculus* and *Aulophorus*, for example, are known to reproduce by the penis breaking into such fragments. This complete regeneration is noteworthy as these Annelid species are the most highly organized animals to have this capability. Many other taxa (such as most earthworms) cannot reproduce this way, though they have varying abilities to regrow amputated segments.

Sexual Reproduction

Sexual reproduction allows a species to better adapt to its environment. Some annelida species are hermaphroditic, while others have distinct sexes.

Most polychaete worms are gonochoristic, that is, they have separate males and females and external fertilization. The earliest larval stage, which is lost in some groups, is a ciliated trochophore, similar to those found in other phyla. The animal then begins to develop its segments, one after another, until it reaches its adult size.

Earthworms and other oligochaetes, as well as the leeches, are hermaphroditic and mate periodically throughout the year in favored environmental conditions. They mate by copulation. Two worms which are attracted by each other's secretions lay their bodies together with their heads pointing opposite directions. The fluid is transferred from the male pore to the other worm. Different methods of sperm transference have been observed in different genera, and may involve internal spermathecae (sperm storing chambers) or spermatophores that are attached to the outside of the other worm's body. The clitella lack the free-living ciliated trochophore larvae present in the polychaetes, the embryonic worms developing in a fluid-filled "cocoon" secreted by the clitellum.

Fossil Record

The annelid fossil record is sparse, but a few definite forms are known as early as the Cambrian. Because the creatures have soft bodies, fossilization of a body is an especially rare event. However, a few annelids, such as the living polychaetes in the Serpulidae, secrete calcareous tubes, and such tubes are fairly common as fossils (although these are not necessarily from annelida, as other animal phyla can also secrete tubes). A fossil group has recently been assigned to the annelids, the machaeridians. These forms were polychaetes with rows of dorsal overlapping shell plates. The hard jaws of certain polychaetes, known as scolecodonts, are known from the Ordovician onward, and are common enough to be used for stratigraphic correlation in some cases. The best-preserved and oldest annelid body fossils come from the Cambrian Lagerstätten such as the Burgess Shale of Canada, and the Middle Cambrian strata of the House Range in Utah. The Annelids are also diversely represented in the Pennsylvanian-age Mazon Creek fauna of Illinois. A few small groups have been treated as separate phyla: the Pogonophora and Vestimentifera, now included in the family Siboglinidae, and the Echiura.

Relationships

The arthropods and their kin have long been considered the closest relatives of the annelids, on account of their common segmented structure, giving rise to the grouping of Articulata. However, a number of differences between the two groups suggest this may be convergent evolution. The other major phylum which is of definite relation to the annelids are the molluscs, which share with them the presence of trochophore larvae. Annelids and Molluscs are thus united as the Trochozoa, a taxon more strongly supported by molecular evidence.

Sipuncula, Echiura and Siboglinidae have traditionally been placed in their own phyla, while Clitellata has been considered separated from the polychaete annelids. But recent research indicates that all of them actually belongs within the Polychaete, even if some of these groups have lost their segmentation.

Classes and Subclasses of Annelida

- *Clitellata*
- Oli*gochaeta* - The class Oligochaeta includes the megadriles (earthworms), which are both aquatic and terrestrial, and the microdrile families such as tubificids, which include many marine members as well. As traditionally defined, the Oligochaeta are paraphyletic.
- *Leeches (Hirudinea)* - These include both bloodsucking external parasites and predators of small invertebrates.
- *Acanthobdellidea and Branchiobdella* - small leech-like clitellates formerly included with the Hirudinea.
- *Aphanoneura*
- *Polychaeta* - This is the largest group of annelids and the majority are marine. All segments are identical each with a pair of parapodia. The parapodia are used for swimming, burrowing and the creation of a feeding current.

CHAPTER–8

Foraminifera

INTRODUCTION

The Foraminifera, ("Hole Bearers") or forams for short, are a large group of amoeboid protists with reticulating pseudopods, fine strands of cytoplasm that branch and merge to form a dynamic net. They typically produce a test, or shell, which can have either one or multiple chambers, some becoming quite elaborate in structure. These shells are made of calcium carbonate ($CaCO_3$) or agglutinated sediment particles. About 275,000 species are recognized, both living and fossil. They are usually less than 1 mm in size, but some are much larger, and the largest recorded specimen reached 19 cm.

Although as yet unsupported by morphological correlates, molecular data strongly suggest that Foraminifera are closely related to the Cercozoa and Radiolaria, both of which also include amoeboids with complex shells; these three groups make up the Rhizaria. However, the exact relationships of the forams to the other groups and to one another are still not entirely clear.

MORPHOLOGY OF FORAMINIERA

Foraminifera taxa are classified primarily on the basis of wall composition, test shape, and chamber shape and arrangement. Types of apertures, aperture position, and ornamentation are often diagnostic at the generic or specific level.

A. Chamber shape and arrangement

initial chamber: proloculus

may have one or more subsequent chambers

see handout for other arrangements of chambers

see Haq and Boersma (on reserve) for other chamber shapes and arrangements

B. Types of coiling

evolute: all chambers visible on both sides

involute: majority of previous coils hidden by outer whorl

trochospiral: spiral up the axis of coiling

side showing all whorls: dorsal

side showing outer whorl: ventral

streptospiral: trochospiral coiling in several planes

C. Types of apertures

D. Positions of apertures

E. Ornamentation

FORAMINIFERAL EVOLUTION

I. Paleozoic

A. single chambered, sacks or tubes

B. coiling:

attached-Tolypammina (Silurian)

planispiral- Ammodiscus (Silurian)

high spiral- Turittela (Silurian)

C. septae:

multichambered (Devonian)

D. new wall types

calcareous microgranular (Devonian)

sugary appearance, microgranular

calcareous fibrous (Devonian)

porcelaneous (Late Carbon.-Permian) Miliolidae

E. increasing complexity and size

e.g. Endothyra and Fusulinidae (Carboniferous)

F. trochospiral coiling (Carboniferous)

G. Permian extinctions of 10 families

II. Mesozoic

A. first deep dwelling- Nodosariidae (Triassic)

B. first ornamented- Nodosariidae (Jurassic)

C. first bathyal- Nodosariidae (Jurassic)

D. first planktic- Nodosariidae (Jurassic)

E. major terminal extinction of terminal Cretaceous planktics

III. Cenozoic

A. most modern planktic and benthic groups evolved in Miocene

B. large calcareous forams declined in the Paleogene, now only in the Indo-Pacific tropics

BENTHIC FORAMINIFERAL FUNCTIONAL MORPHOLOGY

Form, Structure, and Environment- "Form follows function", as within other organisms, foraminifera develop similar morphologies in similar environments (Convergent Evolution)

I. Buliminids and Uvigerinids

are well know examples exhibiting different forms in different depth habitats.

species abundance correlates with dissolved oxygen

suggests form controlled by respiration or gas exchange of cell

compressed form: high surface area/volume, better facilitating gas exchange

II. Compressed forms

Exhibited by many macroscopic foraminiera but form "follows a different function"

A. 3 forms: flattened

8% of lituolids (microgranular calcite with complicated interiors) discoidal

7% of miliolids (porcellaneous-imperforate)

8% rotalids

B. planispiral-fusiform

50% fusulinids

10% miliolids

In the above cases the large surface area/volume appears adaptive to increasing the efficiency of symbionts. Many inhabited low-nutrient, tropical waters.

III. Other Form and Habitat Relationships

A. Spherical species and genera: often parasitism

B. An erect test with aperature(s) oriented away from the substrate: suspension feeders

C. Mostly trochoid, lenticular or flattened: active herbivores

27% miliolids

21% fusulinids

19% lituolids

17% nodosarids

52% rotalids

20% of all foram genera

D. Chamberlets: test strengthing with thin walls for symbiont photosynthesis.

FORAMINIFERA AND ANALYSES OF PALEOENVIRONMENT: SOME EXAMPLES

I. Planktic Foraminifera

A. Coiling direction and sea-surface temperature (*Neoglobo-quadrina pachyderma*)

B. Test diameter, salinity, and temperature (*Orbulina universa*)

C. Water mass migrations (*California margin*)

II. Benthic Foraminifera

A. Reef environments (handout)

B. Gulf of Mexico: shelf (handout)

C. Vertical water-mass migrations (Pacific)

STRATIGRAPHY

Standard zones have been established which are commonly refered to in Cenozoic stratigraphic studies. These zones have widespread application in low to mid-latitudes. The zonal scheme for the Neogne was established by Banner and Blow (Banner, F.T., Blow, W.H., 1965, Progress in the planktonic foraminiferal biostratigraphy of the Neogene. Nature, 208:1164-1166.) and subsequent papers established a zonal scheme for the Paleogene.

CALCAREOUS NANNOFOSSILS

I. Introduction to Calcareous Nanno

Calcareous nannoplankton constitute a diverse group of calcareous microfossils which appear to be related to modern (extant) coccolithophores.

Coccolithophores are:

marine algae belonging to the Haptophyceae (algae)

most living species possess a flagellar apparatus called the haptonema

photoautotrophic

sometimes have a motile phase and a non-motile phase which normally secretes coccoliths

II. GEOLOGIC RANGE (includes related ancestors)

first appeared in the earliest Jurassic

first diversification in early part of the late Jurassic (Oxfordian)

very diverse by the end of the Cretaceous

massive extinctions during the terminal Cretaceous

III. STRATIGRAPHY

Since the advent of the Deep Sea Drilling Project the stratigraphic and biogeographic distribution of calcareous nannoplankton have been studied in great detail.

In most latitudes, stratigraphic ranges have been partially calibrated to paleomagnetic stratigraphy.

Diveristy is much reduced at high latitudes reducing their utility in these regions

Diachroneity exists with respect to datums across latitude

Standard zones have been established which are commonly refered to in Cenozoic stratigraphic studies:

Martini and Worsley (1970) established the standard zonal scheme for the Neogene and Quaternary and Martini (1970) for the Paleogene.

the Neogene has 21 zones; NN1 (the oldest) to NN21 (the youngest). Note: NN stands for nannofossils-Neogene.

the Paleogene has 25 zones; NP1 (earliest Paleocene) to NP25 (latest Oligocene). Note: NP stands for nannofossils-Paleogene.

IV. Key Cenozoic Genera Non-coccolithophore Nannoliths

1. *Ceratoliths:* Horse-shoe shapped (e.g. Ceratolithus and Amaurolithus)
2. *Discoasters:* Disc-shaped, rayed asteroliths, with or without stems or knobs. (e.g. Discoaster, Tribrachiatus).
3. *Fasiculithids:* Subcylindrical proximal column of wedge-shaped crystals whose thin edges meet along the centre.

Proximal end may have an apical spine. (One genus-Fasiculithus).

4. *Sphenolithids:* Rounded to polygonal base, mounted by an elongated rib-cone with one or more spines radiating from the center (Sphenolithus)
5. *Heliolithids:* Two abutting unequal shields lacking a connecting tube and one shield with subradial sutures (Heliolithus)

Genera Incertae Sedis

1. *Isthmolithus:* Shaped like an elongate parallelogram with wide walls and transverse septa.
2. *Triquetrorhabdulus:* Rodlike nannoliths with three blades or edges 120o apart and pointed or rounded ends.

Coccolithophores and Related Nannoliths

Coccolithids: Two-shielded coccoliths, each with one or more cycles of lelements connected at their inner margins.

Coccolithus- oval with central pore

Cruciplacolithus- +-shape dcentral structure

Chiasmolithus- x-shaped central structure Reticulofenestra-central area with fine reticulate grill

Rhabdosphaerids A central process that gives the coccolith a funnel or dome-shaped appearance. Rhabdosphaera

V. Paleoenvironmental Application

A. Biogeography

B. Frequency fluctuations of genera and species

LIVING FORAMS

Modern forams are primarily marine, although they can survive in brackish conditions. A few species survive in fresh water and one even lives in damp rainforest soi They are very common in the meiobenthos, and about 40 morphospecies are planktonic. This count may however represent only a fraction of actual diversity, since many genetically discrepant species may be morphologically indistinguishable The cell is divided into granular endoplasm and transparent ectoplasm. The pseudopodial net may emerge through a single opening or many perforations in the test, and characteristically has small granules streaming in both directions.

The pseudopods are used for locomotion, anchoring, and in capturing food, which consists of small organisms such as diatoms or bacteria. A number of forms have unicellular algae as endosymbionts, from diverse lineages such as the green algae, red algae, golden algae, diatoms, and dinoflagellates Some forams are kleptoplastic, retaining chloroplasts from ingested algae to conduct photosynthesis.

The foraminiferal life-cycle involves an alternation between haploid and diploid generations, although they are mostly similar in form. The haploid or gamont[*disambiguation needed*] initially has a single nucleus, and divides to produce numerous gametes, which typically have two flagella. The diploid or schizont is multinucleate, and after meiosis fragments to produce new gamonts. Multiple rounds of asexual reproduction between sexual generations is not uncommon in benthic forms Foramanifera typically live for about a month.

Tests

Fossil nummulitid foraminiferans showing microspheric and megalospheric individuals; Eocene of the United Arab Emirates; scale in mm.

The form and composition of the test is the primary means by which forams are identified and classified Most have calcareous tests, composed of calcium carbonate.[In other forams the test may be composed of organic material, made from small pieces of sediment cemented together (agglutinated), and in one genus of silica. Openings in the test, including those that allow cytoplasm to flow between chambers, are called apertures.

Tests are known as fossils as far back as the Cambrian period and many marine sediments are composed primarily of them. For instance, the limestone that makes up the pyramids of Egypt is composed almost entirely of nummulitic benthic foraminifera.

Production estimates indicate that reef foraminifera annually generate approximately 43 million tons of calcium carbonate and thus play an essential role in the production of reef carbonates.

Genetic studies have identified the naked amoeba "Reticulomyxa" and the peculiar xenophyophores as foraminiferans without tests A few other amoeboids produce reticulose pseudopods, and were formerly classified with the forams as the Granuloreticulosa, but this is no longer considered a natural group, and most are now placed among the Cercozoa.

EVOLUTIONARY SIGNIFICANCE

Dying planktonic foraminifera continuously rain down on the sea floor in vast numbers, their mineralized tests preserved as fossils in the accumulating sediment. Beginning in the 1960s, and largely under the auspices of the Deep Sea Drilling, Ocean Drilling, and International Ocean Drilling Programmes, as well as for the purposes of oil exploration, advanced deep-sea drilling techniques have been bringing up sediment cores bearing foraminifera fossils by the millions. The effectively unlimited supply of these fossil tests and the relatively high-precision age-control models available for cores has produced an exceptionally high-quality planktonic foraminifera fossil record dating back to the mid-Jurassic, and presents an unparalleled record for scientists testing and documenting the evolutionary process. The exceptional quality of the fossil record has allowed an impressively detailed picture of species inter-relationships to be developed on the basis of fossils, in many cases subsequently validated independently through molecular genetic studies on extant specimens.

USES OF FORAMS

Because of their diversity, abundance, and complex morphology, fossil foraminiferal assemblages are useful for biostratigraphy, and can accurately give relative dates to rocks. The oil industry relies heavily on microfossils such as forams to find potential oil deposits.

Calcareous fossil foraminifera are formed from elements found in the ancient seas they lived in. Thus they are very useful in paleoclimatology and paleoceanography. They can be used to reconstruct past climate by examining the stable isotope ratios of oxygen, and the history of the carbon cycle and oceanic productivity by examining the stable isotope ratios of carbon. Geographic patterns seen in the fossil records of planktonic forams are also used to reconstruct ancient ocean currents. Because certain types of foraminifera are found only in certain environments, they can be used to figure out the kind of environment under which ancient marine sediments were deposited.

For the same reasons they make useful biostratigraphic markers, living foraminiferal assemblages have been used as bioindicators in coastal environments, including indicators of coral reef health. Because calcium carbonate is susceptible to dissolution in acidic conditions, foraminifera may be particularly affected by changing climate and ocean acidification.

Foraminifera can also be utilised in archaeology in the provenancing of some stone raw material types. Some stone types, such as chert, are commonly found to contain fossilised foraminifera. The types and concentrations of these fossils within a sample of stone can be used to match that sample to a source known to contain the same 'fossil signature'.

DIATOM

Scientific Classification

Domain : Eukaroyto

Kingdom : Chromalveolata

Phylum : Heterokontophyta

Class : Bacillariophyceae

Order

- Centrales
- Pennales

Diatoms (Greek: *d (dia)* = "through" + *(temneir.)* = "to cut", i.e., "cut in half") are a major group of eukaryotic algae, and are one of the most common types of phytoplankton. Most diatoms are unicellular, although they can exist as colonies in the shape of filaments or ribbons (e.g. *Fragillaria*), fans (*Meridion*), zigzags (*Tabellaria*), or stellate colonies (*Asterionella*). Diatoms are producers. A characteristic feature of diatom cells is that they are encased within a unique cell wall made of silica (hydrated silicon dioxide) called a frustule. These frustules show a wide diversity in form, some quite beautiful and ornate, but usually consist of two asymmetrical sides with a split between them, hence the group name. Fossil evidence suggests that they originated during, or before, the early Jurassic Period. Diatom communities are a popular tool for monitoring environmental conditions, past and present, and are commonly used in studies of water quality.

General Biology

There are more than 200 genera of living diatoms, and it is estimated that there are approximately 100,000 extant species. Diatoms are a widespread group and can be found in the oceans, in freshwater, in soils and on damp surfaces. Most live pelagically in open water, although some live as surface films at the water-sediment interface (benthic), or even under damp atmospheric conditions. They are especially important in oceans, where they are estimated to contribute up to 45% of the total oceanic primary production Although usually microscopic, some species of diatoms can reach up to two millimetres in length.

Diatoms belong to a large group called the heterokonts, including both autotrophs (e.g. golden algae, kelp) and heterotrophs (e.g. water moulds). Their yellowish-brown chloroplasts are typical of heterokonts, with four membranes and containing pigments such as fucoxanthin. Individuals usually lack flagella, but they are present in gametes and have the usual heterokont structure, except they lack the hairs (mastigonemes) characteristic in other groups. Most diatoms are non-motile, although some move via flagellation. As their relatively dense cell walls cause them to readily sink, planktonic forms in open water usually rely on turbulent mixing of the upper layers by the wind to keep them suspended in sunlit surface waters. Some species actively regulate their buoyancy with intracellular lipids to counter sinking.

Diatom cells are contained within a unique silicate (silicic acid) cell wall comprising two separate valves (or shells). The biogenic silica that the cell wall is composed of is synthesised intracellularly by the polymerisation of silicic acid monomers. This material is then extruded to the cell exterior and added to the wall. Diatom cell walls are also called frustules or tests, and their two valves typically overlap one other like the two halves of a petri dish. In most species, when a diatom divides to produce two daughter cells, each cell keeps one of the two valves and grows a smaller valve within it. As a result, after each division cycle the average size of diatom cells in the population gets smaller. Once such cells reach a certain minimum size, rather than simply divide vegetatively, they reverse this decline by forming an auxospore. This expands in size to give rise to a much larger cell, which then returns to size-diminishing divisions. Auxospore production is almost always linked to meiosis and sexual reproduction.

Decomposition and decay of diatoms leads to organic and inorganic (in the form of silicates) sediment, the inorganic component of which can lead to a method of analyzing past marine environments by corings of ocean floors or bay muds, since the inorganic matter is embedded in deposition of clays and silts and forms a permanent geological record of such marine strata.

Classification

Selections from Ernst Haeckel's 1904 *Kunstformen der Natur* (Artforms of Nature), showing pennate (left) and centric (right) frustules.

The classification of heterokonts is still unsettled, and they may be treated as a division (or phylum), kingdom, or something in-between. Accordingly, groups like the diatoms may be ranked anywhere from class (usually called Bacillariophyceae) to division (usually called Bacillariophyta), with corresponding changes in the ranks of their subgroups. The diatoms are also sometimes referred to as Class Diatomophyceae.

Diatoms are traditionally divided into two orders: centric diatoms (Centrales), which are radially symmetric, and pennate diatoms (Pennales), which are bilaterally symmetric. The former are paraphyletic to the latter. A more recent classification divides the diatoms into three classes: centric diatoms (Coscinodiscophyceae), pennate diatoms without a raphe (Fragilariophyceae), and pennate diatoms with a raphe (Bacillariophyceae). It is probable there will be further revisions as understanding of their relationships increases.

Planktonic forms in freshwater and marine environments typically exhibit a "boom and bust" (or "bloom and bust") lifestyle. When conditions in the upper mixed layer (nutrients and light) are favourable (*e.g.* at the start of spring) their competitive edge allows them to quickly dominate phytoplankton communities ("boom" or "bloom"). As such they are often classed as opportunistic r-strategists (*i.e.* those organisms whose ecology is defined by a high growth rate, *r*).

When conditions turn unfavourable, usually upon depletion of nutrients, diatom cells typically increase in sinking rate and exit the upper mixed layer ("bust"). This sinking is induced by either a loss of buoyancy control, the synthesis of mucilage that sticks diatoms cells together, or the production of heavy resting spores. Sinking out of the upper mixed layer removes diatoms from conditions inimical to growth, including grazer populations and higher temperatures (which would otherwise increase cell metabolism). Cells reaching deeper water or the shallow seafloor can then rest until conditions become more favourable again. In the open ocean, many sinking cells are lost to the deep, but refuge populations can persist near the thermocline.

Ultimately, diatom cells in these resting populations re-enter the upper mixed layer when vertical mixing entrains them. In most circumstances, this mixing also replenishes nutrients in the upper mixed layer, setting the scene for the next round of diatom blooms. In the open ocean (away from areas of continuous upwelling), this cycle of bloom, bust, then return to pre-bloom conditions typically occurs over an annual cycle, with diatoms only being prevalent during the spring and early summer. In some locations, however, an autumn bloom may occur, caused by the breakdown of summer stratification and the entrainment of nutrients while light levels are still sufficient for growth. Since vertical mixing is increasing, and light levels are falling as winter approaches, these blooms are smaller and shorter-lived than their spring equivalents.

In the open ocean, the condition that typically causes diatom (spring) blooms to end is a lack of silicon. Unlike other nutrients, this is only a major requirement of diatoms so it is not regenerated in the plankton ecosystem as efficiently as, for instance, nitrogen or phosphorus nutrients. This can be seen in maps of surface nutrient concentrations - as nutrients decline along gradients, silicon is usually the first to be exhausted (followed normally by nitrogen then phosphorus).

Because of this bloom-and-bust lifestyle, diatoms are believed to play a disproportionately important role in the export of carbon from oceanic surface waters. Significantly, they also play a key role in the regulation of the biogeochemical cycle of silicon in the modern ocean.

The use of silicon by diatoms is believed by many researchers to be the key to their ecological success. In a classic study found that diatom dominance of mesocosm communities was directly related to the availability of silicate. When silicon content approaches a concentration of 2 mmol m^{-3}, diatoms typically represent more than 70% of the phytoplankton community. It is noted that, relative to organic cell walls, silica frustules require less energy to synthesize (approximately 8% of a comparable organic wall), potentially a significant saving on the overall cell energy budget. Other researchers have suggested that the biogenic silica in diatom cell walls acts as an effective pH buffering agent, facilitating the conversion of bicarbonate to dissolved CO_2 (which is more readily assimilated). Notwithstanding the possible advantages conferred by silicon, diatoms typically have higher growth rates than other algae of a corresponding size.

Diatoms occur in virtually every environment that contains water. This includes not only oceans, seas, lakes and streams, but also soil.

Evolutionary History

Heterokont chloroplasts appear to be derived from those of red algae, rather than directly from prokaryotes as occurred in plants. This suggests they had a more recent origin than many other algae. However, fossil evidence is scant, and it is really only with the evolution of the diatoms themselves that the heterokonts make a serious impression on the fossil record.

The earliest known fossil diatoms date from the early Jurassic (~185 Ma), although molecular clock and sedimentary evidence suggests an earlier origin. It has been suggested that their origin may be related to the end-Permian mass extinction (~250 Ma), after which many marine niches were opened. The gap between this event and the time that fossil diatoms first appear may indicate a period when diatoms were unsilicified and their evolution was cryptic. Since the advent of silicification, diatoms have made a significant impression on the fossil record, with major deposits of fossil diatoms found as far back as the early Cretaceous, and some rocks (diatomaceous earth, diatomite, kieselguhr) being composed almost entirely of them.

Although the diatoms may have existed since the Triassic, the timing of their ascendancy and "take-over" of the silicon cycle is more recent. Prior to the Phanerozoic (before 544 Ma), it is believed that microbial or inorganic processes weakly regulated the ocean's silicon cycle. Subsequently, the cycle appears dominated (and more strongly regulated) by the radiolarians and siliceous sponges, the former as zooplankton, the latter as sedentary filter feeders primarily on the continental shelves. Within the last 100 My, it is thought that the silicon cycle has come under even tighter control, and that this derives from the ecological ascendancy of the diatoms.

However, the precise timing of the "take-over" is unclear, and different authors have conflicting interpretations of the fossil record. Some evidence, such as the displacement of siliceous sponges from the shelves, suggests that this takeover began in the Cretaceous (146 Ma to 65 Ma), while evidence from radiolarians suggests "take-over" did not begin until the Cenozoic (65 Ma to present). Nevertheless, regardless of the details of the "take-over" timing, it is clear that this most recent revolution has installed much tighter biological control over the biogeochemical cycle of silicon.

Collection

Living diatoms are often found clinging in great numbers to filamentous algae, or forming gelatinous masses on various submerged plants. *Cladophora* is frequently covered with *Cocconeis*, an elliptically shaped diatom; *Vaucheria* is often covered with small forms. Diatoms are frequently present as a brown, slippery coating on submerged stones and sticks, and may be seen to "stream" with river current.

The surface mud of a pond, ditch, or lagoon will almost always yield some diatoms. They can be made to emerge by filling a jar with water and mud, wrapping it in black paper and letting direct sunlight fall on the surface of the water. Within a day, the diatoms will come to the top in a scum and can be isolated.

Since diatoms form an important part of the food of molluscs, tunicates, and fishes, the alimentary tracts of these animals often yield forms that are not easily secured in other ways. Marine diatoms can be collected by direct water sampling, though benthic forms can be secured by scraping barnacles, oyster shells, and other shells.

The silicious shells of diatoms are among the most beautiful objects which can be examined with the microscope. To obtain perfectly clean mounts requires considerable time and patience, but once the material is cleaned, preparations may be made at any time with very little trouble.

Genome Sequencing

The entire genomes of the centric diatom, *Thalassiosira pseudonana* and the pennate diatom, *Phaeodactylum tricornutum*, have been sequenced. The first insights into the genome properties of the *P. tricornutum* gene repertoire was described using 1,000 ESTs. Subsequently, the number of ESTs was extended to 12,000 and the Diatom EST Database was constructed for functional analyses These sequences have been used to make a comparative analysis between *P. tricornutum* and the putative complete proteomes from the green alga *Chlamydomonas reinhardtii*, the red alga *Cyanidioschyzon merolae*, and *T. pseudonana*.

Comparisons of the two fully sequenced diatom genomes finds that, despite relatively recent evolutionary divergence (90 million years), around 40% of their genes are not shared More significantly, the analysis finds that the genomes of both species contain hundreds of genes from bacteria, many of which are shared between the species.

Nanotechnology Research

The deposition of silica by diatoms may also prove to be of utility to nanotechnology. Diatom cells repeatedly and reliably manufacture valves of shapes and sizes, potentially allowing diatoms to manufacture micro- or nano-scale structures which may be of use in a range of "widgets" including: optical systems; semiconductor nanolithography; and even using diatom valves as vehicles for drug delivery. Using an appropriate artificial selection procedure, diatoms that produce valves of particular shapes and sizes could be evolved in the laboratory, and then used in chemostat cultures to mass produce nanoscale components.

CHAPTER–9

Radiolarian

INTRODUCTION

Radiolarians (also radiolaria) are amoeboid protozoa that produce intricate mineral skeletons, typically with a central capsule dividing the cell into inner and outer portions, called endoplasm and ectoplasm. They are found as zooplankton throughout the ocean, and their skeletal remains cover large portions of the ocean bottom as radiolarian ooze. Due to their rapid turn-over of species, they represent an important diagnostic fossil found from the Cambrian onwards. Some common radiolarian fossils include *Actinomma, Heliosphaera* and *Hexadoridium*.

Radiolarians have many needle-like pseudopodia supported by bundles of microtubules, called axopods, which aid in the Radiolarian's buoyancy. The nuclei and most other organelles are in the endoplasm, while the ectoplasm is filled with frothy vacuoles and lipid droplets, keeping them buoyant. Often it also contains symbiotic algae, especially zooxanthellae, which provide most of the cell's energy. Some of this organization is found among the heliozoa, but those lack central capsules and only produce simple scales and spines.

The main class of radiolarians are the Polycystinea, which produce siliceous skeletons. These include the majority of fossils. They also include the Acantharea, which produce skeletons of strontium sulfate. Despite some initial suggestions to the contrary, genetic studies place these two groups close together. They also include the peculiar genus *Sticholonche*, which lacks an internal skeleton and so is usually considered a heliozoan.

Traditionally the radiolarians have also included the Phaeodarea, which produce siliceous skeletons but differ from the polycystines in several other respects. However, on molecular trees they branch with the Cercozoa, a group including various flagellate and amoeboid protists.

The other radiolarians appear near, but outside, the Cercozoa, so the similarity is due to convergent evolution. The radiolarians and Cercozoa are included within a supergroup called the Rhizaria.

Some radiolarians are known for their resemblance to regular polyhedra, such as with the icosohedron-shaped *Circogonia icosahedra* pictured to the left.

Fossil Record

The earliest known radiolaria date to the very start of the Cambrian period, appearing in the same beds as the first small shelly fauna - they may even be terminal precambrian in age. They differ little from later radiolaria.

Haeckel's Radiolarians

German biologist Ernst Haeckel produced exquisite (and perhaps somewhat exaggerated) drawings of radiolaria, helping to popularize these protists among Victorian parlor microscopists alongside foraminifera and diatoms.

AMOEBOID

Amoeboids are unicellular lifeforms that mainly consist of contractile vacuoles, a nucleus, and cytoplasm as their basic structure. They move and feed by means of temporary cytoplasmic projections, called pseudopods (false feet). They have appeared in a number of different groups. Some cells in multicellular animals may be amoeboid, for instance human white blood cells, which consume pathogens. Many protists also exist as individual amoeboid cells, or take such a form at some point in their life-cycle. The most famous such organism is

Amoeba proteus; the name amoeba is variously used to describe its close relatives, other organisms similar to it, or the amoeboids in general.

Morphological Categories

Amoeboids may be divided into several morphological categories based on the form and structure of the pseudopods. Those where the pseudopods are supported by regular arrays of microtubules are called actinopods, and forms where they are not are called rhizopods, further divided into lobose, filose, and reticulose amoebae. There is also a strange group of giant marine amoeboids, the xenophyophores, that do not fall into any of these categories.

- *Lobose Pseudopods*

Lobose pseudopods are blunt, and there may be one or several on a cell, which is usually divided into a layer of clear ectoplasm surrounding more granular endoplasm. Most, including *Amoeba* itself, move by the body mass flowing into an anterior pseudopod. The vast majority form a monophyletic group called the Amoebozoa, which also includes most slime molds. A second group, the Percolozoa, includes protists that can transform between amoeboid and flagellate forms.

- *Filose Pseudopods*

Filose pseudopods are narrow and tapering. The vast majority of filose amoebae, including all those that produce shells, are placed within the Cercozoa together with various flagellates that tend to have amoeboid forms. The naked filose amoebae comprise two other groups, the vampyrellids and nucleariids. The latter appear to be close relatives of animals and fungi.

- *Reticulose pseudopods*

Reticulose pseudopods are cytoplasmic strands that branch and merge to form a net. They are found most notably among the Foraminifera, a large group of marine protists that generally produce multi-chambered shells. There are only a few sorts of naked reticulose amoeboids, notably the gymnophryids, and their relationships are not certain.

- *Actinopods*

Actinopods are divided into the radiolaria and heliozoa. The radiolaria are mostly marine protists with complex internal skeletons, including central capsules that divide the cells into granular endoplasm and frothy ectoplasm that keeps them buoyant. The heliozoa include both freshwater and marine forms that use their axopods to capture small prey, and only have simple scales or spines for skeletal elements. Both groups appear to be polyphyletic.

However, amoeboids have appeared separately in many other groups, including various different lines of algae not listed above.

- *Subphylum Sarcodina*

Sarcodina is a subphylum of the phylum Sarcomastigophora, of unicellular life forms that move by cytoplasmic flow. Some species use cytoplasmic extensions called pseudopodia for locomotion or feeding. The subphylum includes such protozoa as the common amoeba and the Foraminifera and Radiolaria. Most members of the subphylum reproduce asexually through fission, although some reproduce sexually. Sarcodina is sometimes subdivided into two classes - Rhizopoda and Actinopoda.

Microtubule

Microtubules are one of the components of the cytoskeleton. They have a diameter of 25 nm and length varying from 200 nanometers to 25 micrometers. Microtubules serve as structural components within cells and are involved in many cellular processes including mitosis, cytokinesis, and vesicular transport.

Structure

Microtubules are polymers of a- and ß-tubulin dimers. The tubulin dimers polymerize end to end in protofilaments. The protofilaments then bundle in hollow cylindrical filaments. Typically, the protofilaments arrange themselves in an imperfect helix with one turn of the helix containing 13 tubulin dimers each from a different protofilament. The image above illustrates a small section of microtubule, a few aß dimers in length.

Another important feature of microtubule structure is polarity. Tubulin polymerizes end to end with the a subunit of one tubulin dimer contacting the ß subunit of the next. Therefore, in a protofilament, one end will have the a subunit exposed while the other end will have the ß subunit exposed. These ends are designated (-) and (+) respectively. The protofilaments bundle parallel to one another, so in a microtubule, there is one end, the (+) end, with only ß subunits exposed while the other end, the (-) end, only has a subunits exposed.

Organization within Cells

Microtubules are nucleated and organized by the microtubule organizing centers (MTOCs), such as centrosomes and basal bodies. They are part of a structural network (the cytoskeleton) within the cell's cytoplasm, but, in addition to structural support, microtubules take part in many other processes, as well. They are capable of growing and shrinking in order to generate force, and there are also motor proteins that allow organelles to move along the microtubule. A notable structure involving microtubules is the mitotic spindle used by eukaryotic cells to segregate their chromosomes correctly during cell division. Microtubules are also part of the cilia and flagella of eukaryotic cells (prokaryote flagella are entirely different).

Nucleation and Growth

Polymerization of microtubules is nucleated in a microtubule organizing center. Contained within the MTOC is another type of tubulin, ?-tubulin, which is distinct from the a and ß subunits which compose the microtubules themselves. The ?-tubulin combines with several other associated proteins to form a circular structure known as the "?-tubulin ring complex." This complex acts as a scaffold for a/ß tubulin dimers to begin polymerization; it acts as a cap of the (-) end while microtubule growth continues away from the MTOC in the (+) direction.

Dynamic Instability

During polymerization, both the a- and ß-subunits of the tubulin dimer are bound to a molecule of GTP. The GTP bound to a-tubulin is stable, but the GTP bound to ß-tubulin may be hydrolized to GDP shortly after assembly. The kinetics of GDP-tubulin are different from those of GTP-tubulin; GDP-tubulin is prone to depolymerization. A GDP-bound tubulin subunit at the tip of a microtubule will fall off, though a GDP-bound tubulin in the middle of a microtubule cannot spontaneously pop out. Since tubulin adds onto the end of the microtubule only in the GTP-bound state, there is generally a cap of GTP-bound tubulin at the tip of the microtubule, protecting it from disassembly. When hydrolysis catches up to the tip of the microtubule, it begins a rapid depolymerization and shrinkage. This switch from growth to shrinking is called a catastrophe. GTP-bound tubulin can begin adding to the tip of the microtubule again, providing a new cap and protecting the microtubule from shrinking. This is referred to as rescue.

In vivo microtubule dynamics vary considerably. Assembly, disassembly and catastrophe rates depend on which microtubule-associated proteins (MAPs) are present.

Chemical Effects on Microtubule Dynamics

Microtubule dynamics can also be altered by drugs. For example, the taxane drug class (e.g. paclitaxel or docetaxel), used in the treatment of cancer, blocks dynamic instability by stabilizing GDP-bound tubulin in the microtubule. Thus, even when hydrolysis of GTP reaches the tip of the microtubule, there is no depolymerization and the microtubule does not shrink back. Nocodazole and Colchicine have the opposite effect, blocking the polymerization of tubulin into microtubules.

Motor Proteins

In addition to movement generated by the dynamic instability of the microtubule itself, the fibers are substrates along which motor proteins can move. The major microtubule motor proteins are kinesin, which generally moves towards the (+) end of the microtubule, and dynein, which generally moves towards the (-) end.

CHAPTER–10

Proterozoic Fossil Record

INTRODUCTION

Origin of the eukaryotic organisms (including the multicellular animals or Metazoa) is commonly considered to be related to growing oxygen content in the atmosphere up to a level that allows aerobic metabolism. Here it is suggested that oxygenation of the biosphere was not a permissive condition but rather a forcing factor that drove evolution towards the formation of complex biological systems. Growing concentration of free oxygen in conjunction with other geohistorical trends acted to chemically impoverish the ocean and atmosphere and made many of the chemical elements immobile or unavailable for metabolic processes. Of particular importance in this connection was the decreasing concentration in sea water of the heavy metals that demonstrate high catalytic ability and make an active center in many enzymes. Increasing biological complexity and the eukaryotization of the biosphere (origin of the eukaryotic cell, growing role of heterotrophy, increasing biodiversity, rise of multicellular organisms, lengthening of trophic chains, acceleration of biological recycling of the chemical elements, etc. can be considered as an evolutionary response to the geochemical deterioration of the environment.

Recent discoveries of the oldest megascopic eukaryotes, such as spiral Grypania (1.9 Ga), the necklace-like colonial organism of tissue-grade organization Horodyskia (1.5 Ga), vermiform Parmia (about 1.0 Ga) and Sinosabellidites (800 Ma ago) are consisitent with the "molecular clock" models on an early origin of animals; metazoans were, however, confinedtorelativelycoldandwelloxygenatedbasinsbeyond the carbonate belt of the ocean until the end of the Proterozoic. Large and diverse invertebrates of the Vendian Period are known mostly from siliciclastic marine basins. This fauna is characterized by high density of the benthic populations and well established clades both at the diploblastic (e.g., *Phylum trilobozoa*) and triploblastic (e.g., *Phylum proarticulata*) grades of organization as well as some taxa related to the Paleozoic phyla. An organic skeleton preceded the rise of the mineralized skeleton in some metazoan phyla. Low temperature of the habitats inhibited biomineralization. Almost simultaneous development of the phosphatic, carbonate and siliceous skeletons in different metazoan groups at the beginning of the Cambrian Period some 545 Ma ago could be related to the colonization of the warm carbonate basins by the metazoans. An additional factor for the rapid diversificationofthe biomineralized phyla could be the growing length of the trophic chains brought about by the rapidly increasing biodiversity and the need for detoxification at the top of the trophic pyramid. Being the byproduct of detoxification, sclerites and spicules, hard mineralized shells and carapaces immediately became an important factor of morphological evolution and growing biodiversity, as well as the object of intensive selection under the growing pressure of predators. Explosive growth of morphophysiological diversity in metazoans during the Vendian and Cambrian had an enormous impact on evolution of other groups of organisms and on the environment.

The origin of multicellular animals has to be considered in the context of the overall increase in biological complexity during the immense time of geological history. The appearance and ecological expansion of the metazoans on the global ecosystem. This rise of biological complexity poses a fundamental problem, which has been tackled by many scientific disciplines because of the growing understanding that the main metabolic pathways and the existing biodiversity have very deep historical causes. The early fossil record, considered in a geohistorical context, provides evidence of the critical importance of the origin and evolution of complex life forms, and helps to test evolutionary models developed by neontological disciplines. However, biological evolution cannot be understood properly without an appreciation of the numerous geological factors affecting the whole hierarchy of life from metabolic pathways to global cycles of the main biophile elements.

Role of Continental Growth in History of Life

The origin of continental plates must have had an enormous impact on life on the early Earth. Vast, stable continents became a trap for a great volume of sedimentary rocks (and chemical elements), thus removing them from an intense recycling in the biosphere. This factor and the activity of the organisms that inhabited vast epicontinental basins became the main factors driving the chemical evolution of the atmosphere and the oceans.

Continental growth by accretion of smaller fragments into large landmasses went through three major episodes (Lowe, 1994). About 5% of the continental crust was formed 3.3-3.1 billion years ago, 58% was formed later, between 2.7 and 2.3 billion years ago, and 33% of the continental crust was formed 2.1-1.6 billion years ago. Thus, about 96% of the continental crust was in place by the Middle Proterozoic.

This extensive continental crust created an entirely new situation in the ocean ecosystem. Large-scale upwelling became a factor enhancing the recycling of biophilic elements in the biosphere. The vast shallow benthic habitats appearing at that time opened up great ecological opportunities because of their enormous diversification of microenvironments, in contrast to the pelagic realm, which is ecologically more uniform. To appreciate the importance of the shallow marine habitats, one has to remember that in the recent ocean, shallow-water environments (0-200 meter deep) occupy less than 8% of the ocean floor, but the total mass of organisms living there constitutes 82.6% of the total biomass of benthic organisms in the ocean. This disproportion relates to the nutrient supply from the depth of the ocean through upwelling and erosion of the land, as well as to the intense recycling of organic matter in the shallow-water environment.

The intensification of nutrient recycling first of all stimulated the primary production by phytoplankton and phytobenthos (including the stromatolite-forming photosynthesizing cyanobacterial and other prokaryotes), that consumed carbon dioxide and released oxygen into the atmosphere. The removal of carbon dioxide from the geochemical cycle due to the burial of organic matter and the formation of carbonate rocks on vast areas of the continents may have been critical factors for a decreasing greenhouse effect during the Proterozoic.

Already in the Early Proterozoic we see a remarkably small number of the carbonate platforms and the first known extensive glacial event, chronologically coinciding with a maximum accumulation of Banded Iron Formations (BIFs) that fixed enormous amounts of oxygen in the form of iron oxides. Burial of carbon in the sediments of the stable continental plates reduced the greenhouse effect down to a very sensitive balance about 850 Ma ago. From this time onwards, the glacial periods became more or less regular events of geological history.

The best way to feel the difference between the early Archean atmosphere and the present one can be obtained by comparing the composition of volcanic gas and normal air. These two extremes show the main trend in the evolution of the atmosphere over the last 4 Ga. Living organisms form the principal factor in this dramatic change. Photoautotrophs, in particular, are responsible for decreasing the concentration of carbon dioxide and increasing the concentration of free oxygen in the atmosphere. The concentration of these two gases determines the rate of weathering of rocks and the availability of some chemical elements in the biosphere. The rate of weathering essential for the recycling of the biophilic elements in the biosphere has also been influenced by the increasing rates of erosion and sedimentation due to the increasing contrast of the planetary relief and the shift in composition of the major feeding provinces of the sedimentary basins - from dark basic rocks to the less resistant acidic and sedimentary rocks.

There are some other examples as well of geochemical trends and events directly or indirectly related to the life activity of biota. There are so-called "extinct sediments", typical for the early geological history only, such as uranium- and gold-bearing conglomerates, Banded Iron Formations, laminated copper deposits in clastic rocks, Pb-Zn ore in shales and carbonates, sedimentary Mn ore, and abundant phosphorites. Many authors point to a decrease in Mg and an increase of Ca in the composition of carbonate sediments in the Proterozoic-Paleozoic transition, a phenomenon that might be related to a growing biological control over carbonate precipitation in the ocean. In the Precambrian ocean, silica simply precipitated from the seawater and formed layers or nodules in the sediments. The rise of organisms, such as sponges, radiolarians, diatoms, and silicoflagellates,capable of sucking up the dissolved silica from the water and using it in the construction of their skeletons, converted the ocean into an almost silica vacuum; the recent seas are undersaturated with regard to this element. This is just one of many examples of a growing biological control over the geochemical cycles in the biosphere.

The geological record shows that most natural environmental factors (including mutagens) have varied in magnitude throughout the history of life. Thus, the background radiation level changed by nearly an order of magnitude during the last 4 Ga. The role of oxygen as the next mutagenic factor was more complex. The temperature of the environment is an important factor as well. The rates of most chemical reactions decrease three to four times for every 10° C drop in temperature. We know that the primary biosphere was relatively hot, and that it slowly cooled down towards the end of the Proterozoic. If so, this general cooling trend of the biosphere had to affect the rates of both geochemical and biochemical reactions (including the metabolic pathways) and genomic evolution.

These three factors, although quite heterogeneous, have had a strong impact on the evolution of the environment. Thus, the decrease in radiogenic heat has caused a slow decline in the volcanic supply to the ocean of those chemical elements that were in abundance at the dawn of life. This effect was reinforced by the retreat of the Moon, contributing greatly to the energy balance of early Earth. Oxygenation of the ocean and the atmosphere has made many chemical elements unavailable through oxidation for metabolism. The temperature decline reduced the rates of biogeochemical reactions including the main metabolic pathways in organisms, the bio-degradation and recycling of metabolites, as well as weathering processes. This triple effect dramatically reduced the former, chemically rich, basis of life and, in a way, created an entirely new world that demanded more complex and efficient enzymatic properties, as well as new sources of energy. Life had to move towards heterotrophy, in particular, to its simplest forms, such as uptake of dissolved organic compounds and predation.

OXYGEN AND ANIMAL LIFE

The origin of eukaryotes including animals is usually considered to be related to the rising oxygen content of the atmosphere up to a level allowing aerobic metabolism. Indeed, an oxygen concentration of ca. 5% PAL (present atmospheric level) is the minimum needed to sustain the activity of most eukaryotic microbes. Around 50% PAL of oxygen is the minimum required by animals. Most recent models on the evolution of the Earth's atmosphere suggest a dramatic rise in oxygen content ca. 2 Ga ago. It is estimates this change from <0.1% PAL to >15% PAL. Another rise in the oxygenation of the atmosphere of up to >50% PAL by the end of the Proterozoic, an event that caused the Cambrian explosion.

It is challenged these models, suggesting that the oxygen content rose very quickly to the present atmospheric level more than 3.5 Ga ago and stayed at ± 50% PAL since then. Taking into account the main factors influencing the dynamics of the geochemical cycles of oxygen. It is proved that atmospheric oxygen content stayed within a very narrow range of 0.6-2 PAL, and, except for local euxinic basins, the ocean was completely oxygenated 2 Ga ago. The presence of cyanobacterial and eukaryotic biomarkers in rocks 2.7 Ga old is consistent with this, although aerated habitats might have been limited to the topmost part of the water column. Anoxic bottom waters probably persisted well after the deposition of banded iron formations ceased. It is suggested that sulfide, rather than oxygen, was responsible for removing iron from deep ocean water. The sulfur fossil record supports this suggestion, showing that ca. 1.8 Ga ago, the increase in sulfide production was sufficient to precipitate the total flux of iron into the oceans. According to Canfield's scenario, deep-ocean oxygenated waters did not develop until the Neoproterozoic 1.0-0.54 Ga ago in association with a second large oxidation of the surface of the Earth and a number of glacial events accompanied by active ocean ventilation by cold polar currents. The evolution of non-photosynthetic sulfide-oxidizing bacteria coincided with a large shift in isotopic composition of biogenic sedimentary sulfides between 0.64 and 1.05 Ga ago. Both events were probably driven by a rise in atmospheric oxygen concentration higher than 5-18% PAL, a change that may also have triggered the evolution of animals. But even when this hypothesis would be correct, vast shallow-water and well aerated habitats within the photic zone could still harbor a eukaryotic community since the Late Archean.

The oxygenation of the biosphere was not such a permissive factor, but rather a forcing one driving the formation of complex biological systems. The increasing concentration of free oxygen has dramatically reduced the geochemical basis of life by making many chemical elements either immobile or unavailable for their metabolism. The removal of an enormous amount of metals from the biogeochemical cycle is documented in abundant sedimentary ores, including "extinct deposits" common during some exclusively Precambrian periods. This is particularly true with regard to heavy metals (W, Co, Ni, Mo, Fe etc.) with a great catalytic ability and that take an active part in the operation of many enzymes, real stirrer-ups and accelerators of metabolic processes. The impoverishment of the geochemical basis of life was caused, in addition to oxygenation, by a steady decline in ocean and atmospheric temperatures during the Precambrian.

The formation of the eukaryotic cell through symbio-genesis of prokaryotes and the subsequent evolution towards biological complexity were their response to the growing oxygenation of the environment in order to protect their metabolic pathways.

In this respect, the membranes of the organelles in the eukaryotic cell and multicellularity look like two phases of a single process leading to the creation of a semi-isolated, compartmentalized and controllable internal environment. The rise of biological complexity was the only way to the most economical existence within the conditions of a chemical impoverishment of the environment. The progressive development of heterotrophy during the late Proterozoic and the early Paleozoic eons can be considered as a consequence of the same causal chain.

The idea that deficiency in the inorganic and organic nutrients can be the driving force in the evolution of eukaryotes is not new, but it should be reconsidered in the light of new data on the very early rise of eukaryotic metabolism.

Eukaryotization as reflected in Proterozoic fossil record. The oldest known rocks containing signs of life are over 3.8 Ga old metasediments in the Itsaq Gneiss complex of Greenland. Rocks of this age are metamorphosed to such a degree that few body fossils, if any, can be preserved; only isotopic geochemistry can provide evidence of life. Eukaryotic cells (and Cyano-bacteria) appeared before 2.7 Ga, as indicated by hydrocarbon biomarkers that have been preserved in the late Archean rocks of Western Australia. It took at least a billion years of evolution to form a complex eukaryotic cell and its numerous genes required for the cytoskeleton, compartmentalization, cell-cycle control, proteolysis, protein phosphorylation and RNA splicing.

If the oldest filamentous microfossils that resemble recent Cyanobacteria by their shape, size and habitat, do indeed belong to these rather advanced prokaryotes, one can suppose that all the major metabolic pathways have been formed no later than 3.5 Ga ago. This conclusion can be drawn from the positioning of the Cyanobacteria at the top of the molecular phylogenetic tree of the Eubacteria, as based on the sequencing of bacterial ribosomal RNA. Recently, doubts have been expressed with respect to the biogenic nature of the oldest known microfossils and that of the oldest stromatolites. However, the critics are unable to prove that all 3.5 Ga old microfossils and stromatolites of roughly the same age ever described are abiogenic, or that the oldest biogeochemical signals of life are doubtful.

The temporal trend in the increase in cell size in Precambrian microfossils is considered as an indication of progressive eukaryotization. The upper size limit of the prokaryotic cell (conventionally accepted near 60 mk) was passed ca. 1.9 Ga ago. There are, however, some examples of recent giant bacterial cells and very small eukaryotic cells. The lower size limit of eukaryotic cells is ca. 20 um (and, occasionally, may be as low as 1-2 um), and their real history may therefore begin much before 1.9 Ga ago.

In rocks 1.7-1.0 Ga old, eukaryotic microfossils are widespread and well preserved, but their assemblage and global diversities are low and their turnover is slow. Roughly 1.5 Ga ago, eukaryotes already showed a modern cytoskeletal complexity. This may be concluded from the observed ability to remodel their cell morphology dynamically during their lifetime, as well as from a clear capacity for subdividing environments ecologically. The early eukaryotic fossil record shows a rise of rhodophytes over 1200 Ma, of dinoflagellates over 1100 Ma, of chlorophytes over 1000 Ma, and of ciliates and testate amoebae over ca. 750 Ma ago. Observed diversity levels of protistan microfossils increased significantly around1 Ga ago, as did their turnover rates. Molecular phylogenies suggest that this was part of a broader radiation of "higher" eukaryotic phyla.

Throughout the Proterozoic fossil record we see an increasing abundance and diversity of megascopic fossils. Most fossils consist of carbonaceous material and look like dried and compressed hotdogs or raisins. These fossils seem to reflect the rise of multicellularity that could have happened independently many times in different kingdoms. The multicellularity arose at least six times: in animals, in several algal taxa, and in fungi. In some taxa, multicellularity may have been genetically preserved as a selective advantage due to large body size and individual biomass, or because of cell specialization.

The oldest megascopic carbonaceous fossils of uncertain nature appear in rocks 1.8-2.1 Ga old. Some of these fossils may represent flattened remains of microbial colonies or fragments of the bacterial mats, while part of them may belong to early eukaryotes. The oldest known megascopic eukaryotic algae are Grypania from the Lower Proterozoic iron formation in Michigan. Originally, the age of these rocks was estimated to be ca. 2.1 Ga, but later only as 1.85 Ga.

Fossils somewhat resembling cellular structures are extremely rare. Morphologically more complex carbonaceous fossils with some possible anatomical details appear in rocks ca. 1.4 Ga old but they become more complex and diverse after 0.8-0.9 Ga ago. The larger part of these fossils is interpreted as eukaryotic algae, although some are considered as probable colonial prokaryotes, fungi, or even as metazoans. The cellularly preserved microfossils of red, green and brown algae indicate that multicellularity was achieved in these groups by about 1.0 Ga ago. The question is now, where are the first multicellular animals?

An unexpectedly complex life form was recently discovered in old rocks of the Middle Proterozoic age of Montana and Western Australia. The enigmatic bedding-plane impressions, resembling a necklace or string of beads), were initially described from the Belt Supergroup, Montana, U.S.A. and later also from the Manganese Group, Bangemall Supergroup, Western Australia. These megascopic fossils, recently named Horodyskia moniliformis are both prolific and geographically widespread and appear to be stratigraphically narrowly constrained. A detailed study of the taphonomy, paleoecology and morphology of these fossils resulted in the reconstruction of Horodyskia as a colonial benthic eukaryotic organism of tissue-grade organization, most probably, of metazoan nature. Wide cones (zooids) growing upward a maximum of 1.0 cm connected with a stolon. The maximum length of the colony reaches 30 cm. The size of the cones and the spacing between them are constant for each colony. The complex mode of growth of the colony, regular arrangement of the zooids, the absence of branching forms, the rigid wall of the cone, and the strong stolon that even survived some transportation of the colony by water currents, all these features suggest a tissue-grade organization and a well established set of regulatory genes. Horodyskia marks the first occurrence of complex multicellularity in the history of life as well as the earliest evidence that organismal morphology became an essential factor in the evolutionary process.

If the interpretation of Horodyskia as an early metazoan is correct, it almost doubles the time range of the metazoan fossil record. Some associated problematic fossils in Montana and Western Australia suggest a greater diversity of megascopic organisms living 1.5 Ga ago. An important peculiarity of these fossils is that they are preserved in siliciclastic rocks and the same way as the Ediacara fauna of the Vendian Period (600-545 Ma ago).

The problematic discoidal and trace-like fossils from tidal sandstones 1.2 Ga old from southwestern Australia do not show convincing evidence of their metazoan nature. More promising are the macroscopic worm-like organisms discovered in the lower Neoproterozoic deposits in Huainan District, Anhui Province, of the northern China Platform. The age of the rocks is believed to be ca. 740 Ma. These carbonaceous fossils resemble thin dark filaments 15-20 mm long and show a very fine annulation (or segmentation). Known as Protoarenicola and Pararenicola, these fossils were assigned to the oldest metazoans, probably annelid worms and pogonophorans. In addition to the general worm-like form, size, and kind of segmentation, these organisms show a broad, circular aperture or proboscis-like structure on one, presumably anterior, end of their slender body. The opposite (posterior) end is rounded.

In 1986, this discovery was followed by a report by Sun Weiguo and coauthors on even older worm-like fossils of Sinosabellidites from rocks 800-850 Ma old and from the same region. These fossils show a very fine and regular transverse striation, although without any other features of complexity. Both ends of the band-like fossil are rounded, and, if not with a faint striation, this fossil would look like the associated Tawuia, compressed, sausage-like fossils, tentatively interpreted as multicellular eukaryotic algae, or Metaphyta.

A description of the new genus Parmia which increases the diversity of worm-like creatures inhabiting the ocean during the

Late Riphean well before the Ediacara fauna. Over 80 specimens have been collected from a core from a borehole drilled in southern Timan, northeast of the Russian Platform. With a maximum length up to 60 mm and a body width up to 2.5 mm, these organisms show a regular homonomous segmentation (7-11 segments per millimeter of body length). Parmia is interpreted as a probable predecessor of the annelid worms. All geological and paleontological evidence (including the representative microfossil assemblage) indicates that the age of Parmia may be about 1 Ga. Although the exact age of Sinosabellidites and Parmia is yet to be determined, these metazoan fossils are certainly much older than the Ediacara-type invertebrates known from the Vendian Period.

MOLECULAR DATA ON ORIGIN OF METAZOANS

Gene sequencing supplies evidence that the fungi diverged from plants and animals roughly 1.6 Ga ago, which is the oldest molecular dating of animal life. This conclusion is supported by several earlier molecular data, although calibration problems of molecular clocks cast doubts onto both old and more recent datings of metazoan origins. Molecular evidence of an older origin of the metazoan phyla was obtained from analyzing calibrated rates of molecular sequence divergence. Independent estimates obtained from the analysis of seven gene loci suggest that the invertebrates diverged from the chordates about a billion years ago. Protostomes diverged from chordate lineages well before the echinoderms, which suggests a prolonged radiation of animal phyla and an origin of animals no later than 1200 Ma ago.

Additional molecular data support this conclusion. However, the conflict between these molecular models and the traditional perception of the fossil record (such as a Cambrian evolutionary explosion) makes some specialists wish to reconsider the molecular data. Eighteen protein-coding gene loci and estimated that the protostomes (arthropods, annelids, and mollusks) diverged from the deuterostomes (echinoderms and chordates) about 670 Ma ago, and the chordates from the echinoderms about 600 Ma ago. Both estimates are still older than the fossil record indicates, but they are close enough to some traditional paleontological estimates. Ayala believes that the results obtained by Wray and coauthors, with corresponding divergence times of 1,200 and 1,000 Ma ago, are imperfect because they extrapolated from slow-evolving vertebrate rates to the faster-evolving invertebrates, with their higher rates of molecular evolution.

In spite of this criticism, a longer Precambrian history of the metazoans cannot be ruled out. Thus, the similarities between evolutionary explosions documented by the fossil record. A literal perception of the fossil record gives an impression of a greatly accelerated morphological evolution during some time periods such as the Early Cambrian (545 Ma ago) and the Early Tertiary (c. 65 Ma ago). The Early Cambrian was always considered as the period of appearance of animal phyla, and the Early Tertiary was the time of appearance of modern birds and the orders of placental mammals. These apparent "explosions" are questioned, however, by cladistic and biogeographic studies. Both approaches supported by molecular data expose prolonged periods with evolutionary innovations prior to those "explosions", although the fossil record tells us very little about those periods.

Molecular biologists identified genomic properties related to multicellularity, finding very few genes specific to all multicellular species. This suggests that the transition to multicellularity may not have required the evolution of many new genes absent from unicellular organisms. The metabolic pathways necessary for multicellular organization could already have been in existence in unicellular eukaryotes. This means that the evolution from unicellular protozoans to some multicellular animals could have taken a very short time. The transition from a colonial cell to true multicellularity (that is, to a tissue-grade organization) required that a part of the genome responsible for autonomous cell life is switched off. The rise of tissues limited the exchange of genetic information between the cells within the organism and between the multicellular organisms as well. This function was successfully overtaken by sexual reproduction.

The molecular-clock approach is a powerful instrument for revealing cryptic events during biological history. The molecular clock shows ways to uncover the time of origination of metazoans and the time of major bifurcations occurred in the phylogenetic tree. But along with the general methodological difficulties related to problems of homology and analogy that most comparative methods share, two fundamental assumptions concerning molecular clocks remain uncertain: constant rates and their dependence on generation time. Discrepancies can be expected to occur between various techniques and between various classes of molecules. Attempts to compromise or to average extreme molecular data do not look very promising. Each class of biomolecules tells its own story in its own language, which we have yet to learn. It is now realized that the molecular evolution of life is a very complex process influenced by numerous intrinsic and extrinsic factors.

One of the most puzzling questions is why, in spite of an early origin of the eukaryotic organisms 2.7 Ga ago, they did not become dominant life forms in the global ecosystem until the end of the Proterozoic, that is ca. 600 Ma ago. Indeed, during the greater part of the Proterozoic, the benthic habitats of the shallow epiplatform basins were dominated by prokaryotic communities, the most widespread of which were the stromatolites.

Stromatolites are biosedimentary structures generated by benthic microbial communities such as bacterial mats and biofilms. They look like mounds, cones, columns, or branching tree-like forms built of thin laminated carbonate rock (simple forms of stromatolites are known from siliciclastic basins as well). The size of these structures ranges from a few millimeters to a few meters high. Microbial mats that today are common in hypersaline marches, lagoons and sabkhas, are considered as stromatolite analogues and as relict ecosystems that once dominated the subaqueous photic realm, or even deeper waters and that were probably widespread on land in lakes throughout most of the Precambrian.

Stromatolites appeared as early as 3.5 Ga ago and became widespread during the Early Proterozoic. The abundance and morphological diversity of the stromatolites increased rapidly during the Late Archean and the Proterozoic and reached their maximum at about 1.0 Ga ago. During the Proterozoic, densely packed and vertically oriented stromatolites covered the bottom of the epiplatform basins and the continental shelf. They formed carbonate platforms, ramps and reefs that seemed to be far more extensive than reefs built by algae and corals during the Phanerozoic. The tremendous morphological diversity of the Precambrian stromatolites and their peculiar micro-structures may reflect the interplay between such factors as the composition of the microbial communities and their environment.

Recent microbial mats do not leave much ecological space for most eukaryotic organisms. Some protists and animals do live in, on and amongst the stromatolites, in spite of the hypersalinity of the water, like the community of Shark Bay, Western Australia, but the diversity of these eukaryotic organisms is very low compared to that in the marine environment with a normal salinity, and some invertebrates are represented by dwarf species. A low diversity of eukaryotes is also observed in fresh-water ecosystems dominated by Cyanobacteria. Recent studies of Cyanobacteria revealed a diversity of toxins (neurotoxins, hepatotoxins, cytotoxins, and several others), which can be directly lethal to small zooplankton and metazoans, or can reduce the size and number of offspring produced by organisms feeding on Cyanobacteria. These facts bring us back to Vologdin's hypothesis that Cyanobacteria responsible for stromatolite formation during the Precambrian might, like their recent counterparts, have produced toxins that inhibited early animal evolution. Bacterial toxins as well as marginal marine environments with a high salinity and a fluctuating humidity seem to be major factors controlling the distribution of metazoan grazers at present utilizing Cyanobacteria, and probably to a larger extent also in the past.

This ecological antagonism of stromatolites and animals was documented in the Phanerozoic fossil record. For instance, the term "disaster forms" for stromatolites and some other microbial structures found in the Lower Triassic strata that dramatically increased in abundance during the recovery interval after a mass extinction event. This fact and observations made in the recent habitats dominated by stromatolites are consistent with the hypothesis that the primary biotope of animals might have existed beyond the prokaryotes-dominated environments.

The catastrophic decline in stromatolite diversity and abundance after 850 Ma ago has been interpreted as indirect evidence of metazoan activity. Some authors suggested that grazing and bioturbation by early invertebrates might have caused the decline of stromatolites during the Late Proterozoic. However, direct paleontological evidence has never supported this idea, and it seems to be inconsistent with recent observations on the ecology of bacterial mats as well. Arthropods are dominant grazers on recent microbial mats, with a minor role for gastropods and fishes, taxa that did not appear as late as 850 Ma ago. Most grazers have a small body size of less than a few millimeters, which is why they do not prevent bacterial mats from developing, but rather coexist with bacterial communities. All these observations cast doubts on the idea that metazoan grazers would have wiped out most Proterozoic stromatolites.

The ecological antagonism between bacterial ecosystems and eukaryotes, which is so obvious at present, may have been even stronger in the Precambrian preventing an early domination of the eukaryotes. That is why it is difficult to accept the idea of a biomat-related lifestyle. Anoxia beneath the bacterial mat because of the decomposing dead organic material would not allow the metazoans to colonize the space below the mat. Contrast change in oxygen saturation at the top millimeter of the cyanobacterial mat (from 100% in daylight down to zero in the night because of the H_2S rising up) excludes most of the eukaryotes from this microenvironment. There is some evidence that the biotopes of the stable bacterial mats (including stromatolites) and the early metazoans were essentially different.

The decline of the stromatolite communities after an almost 2 Ga long global dominance in the benthic realm may have been caused by a combination of heterogeneous factors, such as the appearance of eukaryotic algae that competed with cyanobacteria for nutrients, habitats and light, extremely low seawater levels (in particular, during the glacial periods), negative effects of increasing concentrations of biogenic oxygen on the bacterial stromatolite-building communities, a decreasing carbonate saturation of the sea water during the Late Proterozoic, climate change, particularly during the Neoproterozoic Glacial Era some 850-600 Ma ago with all its paleogeographic and geochemical consequences. In fact, all these hypotheses may complement one another. The positive correlation of water temperatures in the recent oceans with the ratio of bacterial production and the primary production of the phytoplankton suggests that the cooling Neoproterozoic biosphere favored eukaryotic life rather than prokaryotic life. The dramatic decline of the stromatolite ecosystems as the ecological dominants of the warm marine basins after 850 Ma ago in fact meant their replacement by eukaryote-dominated ecosystems. This process opened the ecospace for the calcareous and other biomineralizing organisms with a high potential of preservation, the main object of classical paleontology.

Climatic Trend during the Proterozoic

The climatic history of the earth does not show any glaciations during most of the Archean eon, rare glacial episodes being found only in the interval from the Late Archean to the Middle Proterozoic, and frequent and periodical glaciations occurring since the late Proterozoic through the Phanerozoic. A cooling mantle, a decrease in radiogenic heat, a lowering intensity of the gravitational differentiation of the mineral substance and of tidal deformations in the subcrustal zones of the Earth are among the factors that will have caused the steady shift from a warm biosphere to a colder one. However, the main causes of cooling were the decrease in concentration of greenhouse gases in the atmosphere and a growing albedo of the planet due to continental growth. The first extensive glaciation on the planet is documented to have fallen some 2.3 Ga ago, after which there was a long period of warm climates until ca. 800 Ma ago. Since then, glacial periods became more or less regular events in geological history.

Between 800 and 550 Ma ago, the biosphere went through an extremely cold period known as the Snowball Earth period with at least four (or perhaps five) successive, severe glaciations. The most extensive glaciation was the Varanger glaciation some 650-620 Ma ago, that immediately preceded the first mass appearance and global expansion of animals in the history of the biosphere. The magnitude of this glacial has been inferred from the extensive geographic distribution of tillites, varves and other cold-related evidence from deposits of this age. The proponents of an extreme climatic model suggest that an ice sheet over a mile thick covered the entire planet from the poles to the equator. According to various estimates, this global ice sheet could have stayed in place for a very long time, from 4 up to 35 million years, having tremendous consequences for life. Banded iron formations associating with glaciogenic rocks are believed to reappear after an absence of 1 Ga long when the ocean, sealed by ice, became anoxic and therefore rich in dissolved ferrous iron. The carbon isotope record shows enormous negative shifts associated with glacial periods, reflecting dramatic changes in the global ecosystem, as well as a presumed collapse of the biological productivity at the ocean surface for millions of years.

This extreme Snowball model is increasingly criticized. No consensus exists about how many extensive glaciations did occur on Earth between 850 and 550 Ma ago. Thus the only two glacial periods occurred during that time, the Sturtian and the Marinoan (Varanger) glaciations, both found before 600 Ma and separated by a period of 100 to 150 Ma. This makes them independent events rather than linked ones according to the proposed mechanism of a "Snowball Earth". A softer scenario for the Cryogenian with an unfrozen band of tropical ocean waters, providing a refugium for the early metazoans. The continuity of the fossil record of eukaryotic organisms (including abundant phytoplankton) throughout the late Proterozoic shows that global ice shields never interrupted photosynthesis, nor eukaryotic life in general.

Remarkably, the time interval of the Cryogenian embraces the minimum age of ca. 700-750 Ma for the first metazoans as predicted by some molecular clock models. Whether we accept this age or chose models that suggest deeper roots for the metazoans, the essential part of early animal life evolved in a cold ocean. In this respect, it is important to know the environmental factors that challenged the oldest metazoans.

Amidst Glacial Period

Normally, glacial periods are accompanied by strong environmental shifts including radical geographic changes. A shield of ice and snow covered many, though not all, Vendian continents and permafrost was widespread over vast areas of land. The equatorial continents free of ice contained dry deserts. Water erosion of the land and nutrient input into the ocean hardly occurred. Life on frozen continents was reduced to only a few habitats such as hydrotherms and the subsurface realm where liquid ground water could be found. Marine environments were affected by vast shelf glacials and by millions of cubic kilometers of floating ice. The sea level dropped 250-300 meters and strongly fluctuated during the few million years of each glacial period. Most of the shelf was exposed to the air (or was covered with ice) and the area of shallow-water habitats was reduced to a small strip along the edge of continental platforms. Along with the cold climate, vast temperature gradients, changes in atmospheric circulation, and an increasing storm frequency were found. Thus, the glaciations were accompanied by glacioeustatic dropping sea levels, regressions of the sea, decreasing surface areas of the shelf, increasing terrestrial surface areas, a radical shrinkage of benthic environments, and a shift of life to the pelagic realm. The geographic isolation of species had increased because of the growing surface area of land and because of climatic differentiation. The low content of biophilic elements (nutrients) in the open ocean may have been compensated partly by the input of metabolites from the deeper ocean zone: heavy cold water was sinking, pushing up bottom water rich in dissolved nutrients. The same mechanism caused the "ventilation" of the ocean: sinking waters brought oxygen into the deeper parts of the ocean.

All these phenomena must have affected the biota. The most negative effect of these glaciations will have been the destruction of the shallow-water habitats of the benthos of the shelf that in recent oceans normally supports over 80% of the benthic biomass. During the Varanger glaciation, marine life was concentrated in the nutrient-poor and ecologically homogeneous pelagic waters. Apart from this, a large part of the phytoplankton, being shut off from sunlight by oceanic ice cover, was eliminated. One may presume that organisms living around hot vents in the deep ocean could have survived these glacial periods. But the low oxygen content (if not anoxia) of the deep benthic realm may not have allowed the metazoans to colonize these environments untill the very end of the Vendian, or even later.

The Neoproterozoic glacial periods beginning ca. 850 Ma ago seemed to have removed the environment from temperatures optimal for prokaryotes, but they may have had advantages for eukaryotic organisms. This formed a prelude to the Vendian radiation of those metazoans that had been able to survive under the severe glacial conditions. The global distribution of the Vendian fauna, together with the nature of the fossiliferous rocks, are evidence of its marine and relatively cold-water environment. The Vendian fauna, and certainly, its predecessors, will have faced all advantages and disadvantages of cold water habitats. It would be worthwhile to look at recent cold-water marine habitats to see the whole spectrum of the environmental factors that might have affected early metazoan evolution.

Recent Marine Cold-water Environments

The extensive literature on cold-water ecosystems of the recent ocean, particularly the Antarctic basins, reveals some peculiarities possibly resembling the Vendian basins that the Ediacara-type fauna inhabited. The recent cold-water communities show a profound difference from tropical ones. These differences may hint at the nature of the environment at the dawn of animal life, in particular during the glacial periods of the Cryogenian between 800 ans 600 Ma ago.

The ecological conditions and requirements of recent cold-water basins compared to those of the temperate and equatorial zones of the World Ocean can be summarized in the following tables.

ECOLOGICAL CONDITIONS

Abiotic environmental characteristics of recent cold-water basins includes:

1. extreme climates and temperature fluctuations;
2. siliciclastic sedimentation without carbonates;
3. intense vertical water circulation;
4. high concentration of phosphates, nitrates and other metabolites;
5. better aeration of water;
6. circumpolar basins;
7. high concentration of organic nutrients;
8. *seasonality:* variability of light regime above the polar (Arctic and Antarctic) circles alternates between continuous darkness in winter and continuous daylight during summer;
9. extensive cloud cover;
10. less transparent water (shorter photic zone);
11. higher water viscosity:
    (a) making it more difficult for filter feeders to pump sea water in and out;
    (b) allowing small particles to float longer; and
    (c) affecting larval and plankton transport (vertical migration).
12. greater gas solubility of cold water (seawater at zero temperature contains ca. 1.6 times more oxygen than at 20 °C, the complexity of CO_2 -carbonate-bicarbonate equilibria prevents us from making a similar comparison for the solubility of carbon dioxide);
13. in cold water, the solubility of biogenic calcite and aragonite is higher than in warm water.

ECOLOGICAL FEATURES

1. The poor terrestrial life on continents contrasts with the abundant life in oceans, the productivity of the terrestrial ecosystems is low both in relative and absolute terms;
2. high total biomass of living organisms, especially phytoplankton;
3. higher bioproduction of primary producers;
4. unique short food chains (phytoplankton-krill as main herbivore-seals and whales as higher predators), another food web consists of a very rich and diverse benthic fauna supported by a seasonal detritus rain from the water column above;
5. polar ecosystems are less diverse than tropical ones;
6. low stability of biocoenoses;
7. dominance of herbivorous planktotrophs in planktonic metazoans;
8. low proportion of predators in the plankton which, during some seasons, can show population explosions;
9. a high proportion of coelenterates (medusae and ctenophores) among the predators;
10. dominance of forms with direct development (without a pelagic larval stage);
11. low growth rate;
12. long lifespan;
13. low generation turnover;
14. seasonal resource limitation;
15. low reproductive potential, with few, relatively large and yolk-rich eggs;
16. high incidence of direct development, brooding, and/or vivipary;
17. slow gametogenesis;

18. semelparous reproduction;
19. late reproductive maturity;
20. indeterminate growth and high longevity;
21. slow embryonic development;
22. reduced gonadal volume;
23. low basal metabolic rates with regard to oxygen uptake;
24. low activity levels, lethargy, and vertical orientation;
25. low dispersal rates;
26. low colonization rates;
27. low population densities;
28. low mortality because of low predation pressure;
29. K-strategy species dominating;
30. density dominance of some species;
31. high nutrient concentration at the surface;
32. high neritic primary production;
33. high benthic biomass due to delayed maturation and great longevity;
34. soft bottom sediment supports rich infaunal communities dominated by lamellibranch mollusks, actinarians, scleractinian corals and holothurians;
35. stability of the physical environment coupled with a low growth rate tend to evolution of efficient, stable community structure with low annual turnover rate (this type of community supports a higher biomass or alternatively, a higher number of individuals per unit of organic matter than the less stable, less efficientassemblagescharacteristic of physically fluctuatingenvironments);
36. rate of microbial carbon degradation through enzyme-catalyzed reactions is temperature-sensitive and should be very low in cold-water environments;

37. polar submergence of shallow-water genera (including eye-bearing species that are commonly known from the photic zone of the intertidal and continental shelf in the temperate-tropical environment; in the polar regions the species belonging to these genera tend to penetrate to the abyssal depth);
38. polar emergence of abyssal genera (including the genera which are known only from the aphotic abyssal depths in the temperate-tropical marine environments);
39. microbes occur both in water and in sediment but their significance in cold marine ecosystem is unknown (that is extremely low!).

PHYSIOLOGICAL FEATURES OF COLD-WATER METAZOANS

Antarctic poikilotherm animals share several features, such as:

1. slow and seasonal growth;
2. delayed maturation;
3. great longevity;
4. large body size (gigantism in pycnogonids, isopods, sponges, amphipods, free-living nematodes), that reduces potential predation and increase individual fecundity, enhancing population recruitment;
5. low overall fecundity;
6. large, yolky eggs;
7. non-pelagic larval development;
8. seasonal reproduction;
9. low basal metabolic rate;
10. reduced physical activity;
11. seasonal response of growth rates, development and reproduction cycle to seasonal environment;
12. advanced newly hatched juvenile stages;

13. asexuals also tend to occur at high altitudes, and in marginal, nutrient-poor environments compared to their close relatives and sexual species;
14. deferred maturity;
15. energy cost of thermodynamic work, necessary for removing calcium carbonate from sea water for constructing skeletal carbonate, increases in cold water;
16. enzyme-catalyzed reactions proceed more slowly at low temperature;
17. pH of body fluids increases with low temperature;
18. periodical starvation resulting in body mass decrease causes reduction of endogenous respiration sometimes by 99% which allows the cell to conserve carbon and energy, and stabilizes viability in a very small part of the initial population;
19. higher viscosity of body fluids affects some physiological and biomechanical aspects of the activity of the life of invertebrates.

MORPHOLOGICAL FEATURES OF COLD-WATER METAZOANS

1. Large number of aberrant morphologies;
2. large size (gigantism in pycnogonids, isopods, sponges, amphipods, free-living nematodes), which reduces potential predation and increases individual fecundity, enhancing population recruitment;
3. dominance of attached forms among benthonic invertebrates;
4. dominance of soft-bodied forms in benthic communities;
5. larger body size than in the forms of the same species living in warm water. It give six mechanisms for attaining and maintaining a large body size that recur in shallow Antarctic biotas:
    1. the elongation of appendages and sense organs;

2. extreme flattening;
3. a lattice construction of slightly mineralized skeleton;
4. stalk elongation, especially common in the passive suspension-feeding organisms;
5. agglutination; and
6. big-bag construction.

Not all of the characteristics mentioned above can be identified in the geological and fossil record of the Neoproterozoic, and even more, not all of them could have existed during a time as remote from the present day as 600-800 Ma. However, with these data in mind, we may get a deeper insight into the early history of animal life.

GENERAL CHARACTERISTICS OF VENDIAN FAUNA

Remarkably many of the features typical for recent cold-water invertebrates can be recognized in the Precambrian metazoans and their habitats, in particular:

1. large body size or even gigantism (when compared with the small, shelly fossils of the Cambrian);
2. dominance of soft-bodied forms;
3. extremely rare organisms with a mineralized skeleton (observed in carbonate basins only);
4. a weak sclerotization of the cover tissues;
5. a low biotic diversity;
6. a high proportion of body plans considered as unusual (or aberrant) in the later fossil record;
7. dominance of sedentary and benthic vagile life forms (which may actually have a taphonomic cause);
8. small percentage of infaunal metazoans, active filterers and scavengers, which may partially be explained by the difficulty of performing a high level of physical activity in cold-water conditions;
9. dominance of seston feeding sedimentators, microphagous and detrivorous organisms;

10. main macrophagous predators are coelenterate polyps (sea anemones) and medusae;
11. no indication of other predators (no bite marks or regeneration structures in the body of fossils which may indicate a low predator activity);
12. maximum population density in the upper sublittoral zone;
13. very short trophic chains;
14. high portion of cosmopolitan species;
15. low diversity at the species level;
16. dominance of the coelenterate grade or diploblastic forms over triploblastic organisms both in number of species and in the number of individuals;
17. high population density of only a few species;
18. vulnerability of communities (reflected in the very short, though recurring, intervals of the stratigraphic distribution of fossil assemblages, which could often be explained by a high seasonal mortality and a low recolonization rate);
19. mass preservation of the Vendian fauna is typically caused by storms that at present are more frequent in the cold-water zones than in the temperate and tropical zones of the ocean;
20. many of the Vendian animals show features of impulse growth (concentric zones, modularity, segments and other characters of serial homology) that may be related to seasonal variability in the light regime, temperature, bloom of the primary producers, activity and reproduction of algal endosymbionts in the Vendian soft-bodied metazoans.

Low Diversity and Slow Radiation of Vendian Fauna

Although the number of collected specimens of Vendian animals on Earth is nearly 10,000, their taxonomic diversity remains very low. Over 220 fossil species of Vendian animals have been described, but less than half of them have proven to be valid. A few distinct grades and clades can be recognized in the fauna. Paradoxically, sponges, which are expected to be abundant in the oldest communities, are represented by just two taxa Palaeophragmodictia and Ausia. By contrast with Recent coelenterates, which make just two of thirty-five generally recognised phyla of the Animal Kingdom and less than 0.001% of overall species diversity, the Vendian fossil assemblages show a much higher proportion or even domination of the diploblastic organisms and a few well established clades. The most remarkable of the latter is Phylum Trilobozoa which includes genera Tribrachidium, Albumares, Anfesta, Ventodyrus, Skinnera, Vendoconularia and a few other tubular forms, such as Pteridinium, and demonstrates vast morphological diversity, high complexity and physiological and ecological specialization. The most rich clade of the triploblastic grade (or Bilateria) is represented by the Phylum Proarticulata which includes genera Dickinsonia, Yorgia, Vendia, Archaeaspis, Andiva, Ovatoscutum and some other taxa which demonstrate an unusual kind of alternating segmentation described as the symmetry of gliding reflection.

With rare exception, most of the Vendian metazoan genera are represented by a single species. Low diversity at the species level may indicate the stage of initial radiation within most of the known clades and a rather low evolutionary rate during the Vendian Period. The latter assumption can be drawn from the relatively short time interval of the Vendian fossil record which hardly exceeds 30 Ma and from the long time range of many Ediacara taxa.

The absence of a mineralized skeleton may be the cause of the low preservation potential of most of the Pre-cambrian metazoans. However, about 25 forms of trace fossils, which are usually preserved together with imprints of soft-bodied animals, give additional evidence for the idea that both the diversity and the activity of the Vendian metazoans in the benthic realm were low indeed when compared with the Cambrian and later Phanerozoic fossil record. So, the taphonomic and ecological conditions can only partially explain low biodiversity of the Vendian invertebrates, and initial radiation of the macroscopic invertebrate clades seems to be the major cause. We also have always to bear in mind that the overwhelming majority of the recent invertebrates are represented by microscopic forms which have very low preservation potential and virtually no fossil record.

The bulk of the literature dealing with the Antarctic basin shows that slightly more than 3100 animal species have been described from the Antarctic Ocean, while just over 1300 species are known from the Arctic Ocean. Compared with the Vendian, this looks like a lot but compared with the faunal diversity from the tropical zone of the recent ocean it is extremely low. Thus, this low diversity may be a primary feature of the cold-water basins, and at the dawn of animal life, especially during glacial periods, this diversity could have been so low that even minor changes in population structure could have had tremendous evolutionary consequences.

Taphonomic Evidence for Cold-water Nature of Vendian Fauna

Unusual preservation of the Ediacaran fauna (the abundance of fossil soft-bodied organisms) is explained by cumulative effect of many biotic and abiotic factors and is consistent with the model of the Cold Cradle of Animal Life. The functional temperature optima of some enzymes that take part in the bacterial degradation of complex organic compounds are essentially higher than those dominating the recent Antarctic waters. The low rates of biological degradation result in the accumulation of living and dead organic material in sediments at ca. 17gC per m2 per year, that is between 9 and 28% of the primary productivity of the surface waters. These estimates are extremely high compared with those of the temperate zone of the oceanic regions. A practical aspect of this may be found in the prospectiing of cold-water paleobasins for source rocks of hydrocarbons.

That the low temperatures are far more effective inhibitors of decay than anoxia. In cold siliciclastic basins of the Vendian the inhibition of biodegradation could have been even stronger than that for recent counterparts because of the supposed relatively low oxygen concentration in the Late Proterozoic atmosphere and stratification of the paleobasins due to the thermal gradient caused by the constant water turbidity in the absence of active filter-feeders, as well as because of low aeration of the weakly or non-bioturbated sediment. These are additional arguments in favor of the special taphonomic window that was widely opened in the late Proterozoic and closed again with the change in physical and biotic conditions of the biosphere.

COLD CRADLE OF ANIMAL LIFE

The most diverse fossil assemblages of the Vendian metazoans, as well as those even older metazoan fossils (see above), are known from siliciclastic basins. This fact as well as some morphological, paleoecological and taphonomic features of the oldest invertebrates suggests that the cradle of animal life may have been the environments of relatively cold-water basins. The very first metazoans did not appear in warm carbonate basins occupied by stromatolite cyanobacterial communities, but rather beyond the carbonate belt of the planet in siliciclastic basins of the temperate or even the polar zones. Well-oxidized and nutrient-rich environments of the cold basins provided an essential advantage to the animals as aerobic heterotrophs and disfavored their ecological antagonists, prokaryotic communities, just as takes place today.

During the Neoproterozoic glacial periods, the Cryogenian, the ocean could have been a hostile environment. The variability of the light regime, related to the thickness of ice cover, could have affected the primary productivity of the phytoplankton, which could have dropped during winter. One could expect annual periods of starvation in the cold, Vendian water ecosystems as well. These starvation periods resulted in a decrease in the initial population down to small numbers of individuals, in morphological change, and in a reduction of 99% of the endogenous respiration, which allows the cell to conserve carbon and energy.

Relatively high concentration of dissolved oxygen allowed the first metazoans to gain large size and individual mass (that seem to be a common adaptation to the conditions of unstable, seasonal food supply) without the development of complex respiratory systems or efficient blood pigments related to oxygen transport such as haemoglobin. The latter supposition is based on a fascinating example of modern Antarctic ice fish,which lack haemoglobin. The cold climate of the Cryogenian Era of 750-600 Ma ago (the "Snowball Earth") provided heterotrophic aerobic megascopic organisms with a unique opportunity to colonize geographically new and vast habitats.

Vendian Metazoans after Ice Age

The oldest known metazoans of the Vendian Period (550-620 Ma ago) are represented by megascopic body fossils and trace fossils. The species diversity of the Vendian fauna looks very limited although the population densities in the shallow-water environments may have been comparable to recent ones. There is no general consensus regarding the nature and systematic position of the fossils. The Vendian fauna seemed to have included both relict species adapted to essentially different environments of earlier periods of the Neoproterozoic and those taxa ancestral to the Phanerozoic phyla. A number of metazoan body plans and physiologies seemed to have gone extinct very early in the Vendian without leaving recent analogues.

So far, among the more than 30 fossil localities from various parts of the world, the White Sea region in the north of the Russian Platform seems to contain the most representative fossil record of the pre-Cambrian metazoans. Here, the faunal elements also known separately from the Terminal Proterozoic of Newfoundland, Namibia, South Australia and other regions have been discovered forming a succession. The extraordinary preservation, including complex three-dimensional fossils, the abundance and diversity of both body and trace fossils make the White Sea region a true window to the world of the most ancient animals.

The more than 25 years of systematic excavations show:

(a) a growth in taxonomic diversity of body fossils and trace fossils upward in the Vendian succession;

(b) a stepwise pattern in the increase in biodiversity;

(c) a variety of species ranges over time;

(d) a historical change in paleo-faunistic connection of the paleobasin; and

(e) a disappearance of metazoan body fossils and bioturbations in the sediments of the brackish paleobasins. At least six faunal assemblages named after dominant fossils are identifiedin the sequence: Calyptrina-Beltanelloides, Ventogyrus, Inaria, Pteridinium, Charnia, Yorgia-Dickinsonia lissa. Biostratigraphic units (biozones and stages) with a high potential for global correlation can be established. A uranium-lead age of zircon of ca. 555.3 Ma for a volcanic ash at the top of the Charnia fossil assemblage in the sea cliffs of Zymnie Gory suggests a minimum age for the triploblastic metazoans as their bilaterian body fossils and trace fossils occur stratigraphically underneath it. The dynamics in biodiversity seem to have no direct connection with the carbon isotope excursions in the Vendian ocean.

Although the majority of the Vendian body fossils have been interpreted as soft-bodied coelenterates (solitary and colonial polyps and medusas), there is also a large group of complex body fossils that represent a higher, bilaterian or triploblastic grade of organization. Many bilaterian fossils have several features in common, particularly their shield-like shape, their bipolarity, and their complete and homonomous metamerism combined with the symmetry of gliding reflection.This group includes Dickinsonia, Ovatoscutum, Chondroplon, Yorgia, Vendia, Archaeaspis, Andiva and other life forms, collectively assigned to the Phylum Proarticulata. Modes of deformation, of growth pattern and of regeneration marks, although rare, indicate that these fossils have a convex, thin and flexible dorsal carapace composed of an organic, non-mineralized, but slowly degradable substance. The high relief of the imprints proves that the soft body of these animals was thick. The interpretation of these fossils as triploblastic invertebrates is based on some preserved features of their internal anatomy as well as on their trails and tracks found in direct association with body fossils.

Priority of Predation in Animal Evolution

Comparative analysis of the metazoan evolutionary models shows that marine macrophagous consumers of living multicellular attached plants always occupy the terminal position on the phylogenetic tree. Most recent marine herbivores are derived from microphages, detritivores, or predators, and have a post-Paleozoic origin. The pre-bilaterian metazoan clades at the base of the tree, such as Porifera, Cnidaria and Ctenophora often considered as relict phyla of the Proterozoic era, contain carnivores and suspension feeders, but no herbivores. The reason why the Cnidaria and Ctenophora (and almost all other metazoan clades) are all carnivorous is that this way of feeding was the primary feature of their ancestors.

The historical priority of carnivore heterotrophy in animal evolution may be related to the late development of enzymes enabling them to digest plant material. Many metazoan herbivores are themselves unable to digest plant material without the assistance of endosymbiotic bacteria and fungi. Animal proteins can far more easily be decomposed even without special ferments after lysis or denaturation when isolated in the gastric pouch.

The absence of bite marks or regeneration signs in the Vendian metazoans has been interpreted as evidence that before the Cambrian explosion there were no predators of invertebrates with hard parts. Later, the Ediacaran biota was described as having been dominated by organisms that took up dissolved organic compounds or that were photosynthetically assisted by algal endosymbionts. In addition, there were herbivores grazing on bacterial mats. The Vendian biota was thus depicted as a peaceful garden in contrast to the more recent marine biota full of predators.

These interpretations do not seem justified as they ignore important fossil evidence of predation in the Vendian fossil record. The Ediacara-type communities include large solitary polyps (sea anemones) with a voluminous gastric cavity (Nemiana, Bonata), tentaculate medusae (Ediacaria, Hiemalora), and a variety of other forms that were certainly predating actively on each other and on smaller invertebrate and their larvae. The absence of bite marks can easily be explained by their way of feeding: the largest part of the predatory lower invertebrates (cnidarians, ctenophorans, flatworms) cannot leave any predation marks on their prey as they have no hard biting organs like teeth, and they swallow and digest their prey whole. The same must be expected of their Vendian ancestors. The earliest evidence of teeth-like structures in metazoans is some comb-like chitinous fossil Redkinia from the Vendian of the Russian Platform. The long spikes show a size gradient along the base (jaw?) and tiny regular spikes between the long ones. This structure could be part of a filtering apparatus which served to catch and retain prey rather than bite pieces from it.

There is no evidence supporting photoendosymbiosis or the utilization of dissolved organic material, although those trophic methods could have been present in the Vendian invertebrates. Modern cnidarian-dominated communities provide an analogue for understanding the trophic roles of ancient gelatinous organisms. Anemones capture and devour medusae, as well as many other animals and plants. Medusae eat everything from minute to large plankton and nekton (including vertebrates) and, in some cases, probe the benthos or ingest detritus. In addition, they commonly prey on one another, some selectively. The large variety of life forms in the Vendian metazoans indicates that they too were trophically structured, much like similar modern biotas dominated by anemones and medusae. Predation was likely very common right from the very beginning of animal life, and, in a way, may have created the metazoans.

Vendian body fossils show widespread modular growth patterns. Many metazoan taxa from the Vendian have a bipolar bilateral shape and a regular serial homology. However, this repetition of homologous parts along the axis of their body cannot be classified as true segmentation (or metamerism sensu stricto) because of the unusual arrangement of the left and right half-segments of the body. Instead of being placed opposite to each other, these body parts alternate. In geometric terms, we can describe this pattern as a symmetry of gliding reflection, instead of a mirror symmetry that is commonest among the metazoans. These unusual segmented bilateral forms are united in the *Phylum proarticulata.*

Oligomeric homonomic forms are less abundant, and nonsegmented forms are extremely rare. There are no examples of heteronomic segmentation among the Vendian Bilateria. Although true metamerism can be suspected to occur in some Vendian bilaterians (for instance, in arthropod-like animals), it is hard to prove it for most segmented forms.

The absolute dominance of segmented forms among the *Vendian bilateria* can indicate that the evolutionary development of bilateral symmetry and metamerism in many lineages of the earliest Metazoa might have been related processes although this did not always lead to coelomates. An analysis of the promorphology of the Vendian invertebrates, carried out by the author, shows evidence in favor of "cyclomeric" models that derive metamerism from the cyclomerism (or antimerism) of the Coelenterata. However, this analysis reveals essentially new aspects that have not been part of theoretical predictions by neontologists.

The symmetry of gliding reflectionintheVendian metazoans may result from a spiral way of growth. In recent eukaryotic organisms, this symmetry related to a spiral way of growing is especially common among plants. The bilateral pattern and offset (alternating) arrangement of the leaves on the stem can be derived from a spiral shift of the growth point along the axis: a signal to grow is given when the growth point gets to the lateral side of the stem. In some colonial metazoans, such as hydrozoan and anthozoan polyps, the offset pattern of the colony is common as well. We can assume that an arrangement of the leaves in plants and polyp branches in cnidarian colonies may be a morpho-physiological adaptation to an environmental gradient related to light direction for plants and to water current for the polyps. But what about solitary bilateral creeping forms such as Yorgia? A hypothesis on the origin of the Eumetazoa from a colonial organism deserves attention and further development in the light of the Vendian fossil record.

An alternative hypothesis here suggested may hold that the spiral cleavage of the egg could develop directly in a spiral way of growth at one pole of the embryo and subsequently in an adult organism. An epibenthic lifestyle with the body axis oriented parallel to the bottom surface seemed to promote the transformation of spiral into bilateral growth with the characteristic offset (alternating) pattern of the homologous parts. This hypothesis may be tested by the observations of the embryonic development of recent invertebrates.

The unusual growth patterns among Vendian bilaterians, such as the Proarticulata, may reveal unknown steps in the early evolution of the metazoan genome particularly, in the Homeobox gene complex. The author has suggested earlier that the genetic control of serial homology or "segmentation" among presumably diploblastic Vendian animals may be similar to that found in bilaterians. The discovery of Hox-type gene clusters within crown diploblasts certainly lends credence to this hypothesis. The serial homology of many Vendian bilaterians suggests a compartmentalization of gene expression events, commonly found in Hox-type gene regulation in metazoans through his-regulatory elements. Recent advances in developmental biology provide convincing evidence of a causal link between the activation of Hox-genes and segmentation. Analysis of the body plan of Vendian fossils may give clues to the evolution of Hox-type genes, and, conversely, advances in developmental genetics may help interpret the most ancient metazoan fossil record.

Alternating left and right segments and asymmetry of the anterior part, which is so obvious in Yorgia and Archaeaspis, remind one of the offset arrangement of the muscle myomeres on the lateral side of modern cephalochordate Amphioxus. This similarity returns us to an old theory by Geoffroy Saint-Hilaire on the origin of the chordates from the arthropod-like invertebrates via inversion of the dorsal side to the ventral one. Since the first half of the19th century this theory experienced acceptance and rejection, and at present new data revive interest in this theory. Presence of chordates (Pikaia and Yunnanozoon) in the Cambrian shows that this group might have originated in the Vendian, and the Proarticulata might be considered among the possible ancestors of the chordates.

A careful study of trace fossils provides information on the activity of metazoan life and environmental variables. The mode of locomotion, feeding habits, behavioral patterns, body morphology, taxis sensitivity, and some physiological functions can all be derived from bioturbations. Normally, it is rather difficult to identify the nature of the producer unless it is a kind of simple dwelling, such as the burrow of a sea anemone. However, our knowledge on the oldest metazoans is too limited to ignore this group of fossils.

Among over 25 forms of Vendian bioturbations, grazing and crawling trails, feeding and dwelling burrows (in order of descending abundance) form the oldest assemblages of trace fossils. Basically "two-dimensional" horizontal behavioral stereotypes dominate, particularly meandering patterns, although vertically oriented dwelling burrows and locomotion tunnels that cross the sediment-water interface are known as well. Normally, penetration into sediment did not go 5 cm below the bottom surface that, along with a limited degree of biological processing of the sediment, might be related to a limited oxidation of the sediment, even in relatively shallow-water habitats. The decreasing diversity and extent of bioturbations in the offshore sedimentary facies may reflect a limited aeration of the deeper benthic realm as well. On average, the body size of the burrowing infaunal animals in the Vendian was less than that of their Cambrian counterparts. A dramatic increase in size, diversity and depth of the bioturbations has been documented for the very end of the Vendian, that is, for the Rovno Regional Stage of the Russian Platform, or for the Nemakit-Daldyn Horizon of the north Siberian Platform.

Peristaltic modes of locomotion by means of a pedal wave seem to have dominated over crawling and burrowing by means of a hydrostatic skeleton or by that of the peristaltics of the whole body. Its proportional occurrence may reflect the evolution of the locomotory diversity from mucociliary creeping to pedal peristaltic waves and further to whole body peristaltics, to wave-like lateral body bending and the usage of appendages. The two latter kinds of locomotion, common in recent invertebrates are not yet documented for the Vendian.

Until very recently, one could not find any correlation between the body fossils and the trace fossils of the Vendian fauna. This situation has changed after a series of important discoveries of locomotion trails and tracks preserved in direct association with Ediacara-type body fossils in Vendian siliciclastic deposits in the White Sea Region, north of the Russian Platform. These fossil associations were formed under the conditions of fast sedimentation of fine sand lying on top of a muddy bottom plane densely inhabited by a benthic fauna. A sedimentation event stopped the activity of scavengers and bioturbators and thus promoted the preservation of body imprints and trace fossils. Probably, due to bacterial mucilage, the muddy bottom was dense enough to retain its primary microrelief under the conditions of stronger water currents. The sedimentation event did not destroy the trails because the mucus produced by the animals glued and sealed the bottom sediment. Fossil animals of two species are especially common at the end of their trails, Yorgia and Kimberella.

Yorgia waggoneri is a large (up to 25 cm) segmented bilateral organism of the triploblastic grade of organization belonging to the Family Dickinsoniidae, Class Dipleurozoa, Phylum Proarticulata. Its trail consists of a series of similar elements ("platforms") that repeat the oval shape of the body and segmented pattern of the ventral side. Maximum observed length of the trail is 4.3 metres. The "platforms" are arranged as a chain or a cluster and they may overlap each other. Being the cast of the ventral surface of the animal, each element of the trail contains the finest details of external morphology of the animal. The smooth surface of the trail, its preservation in a positive relief on the sole of the sandstone bed, absence of a sharp borderline between the trail and the irregular microrelief of the surrounding bedding plane and the uniform morphology of the "platforms" make a fundamental difference between the ichnofossil and the associated body imprint. Body fossils normally show a broad spectrum of deformations and preservation forms in a deep negative hyporelief (concave imprint on the sole of the sandstone bed).

Whilst crawling or gliding over the mud and staying in a resting phase, Yorgia produced abundant mucus similar to recent flat worms or gastropods. This mucus impregnated, glued and sealed the sediment and thus protected the trail from erosion during the sedimentation event. Short pulses of locomotion (gliding over the bottom surface) alternated with long resting phases, during which more mucus was produced and an exact imprint was fixed of the ventral side in the sediment. This study of Yorgia and its locomotion trails directs our attention to some other segmented fossils preserved in positive relief, particularly those documented in the same bedding plane of the specimens just mentioned. These long segmented fossils, preserved in a very low positive relief may well be candidates for the locomotion trails of another dickinsoniid taxon. This was recently confirmed by the discovery of long crawling trails produced by Dickinsonia tenuis and *D. costata*, as well as by a bilaterian Epibaion (Dickinsoniidae). The body size of Epibaion reaches 45 cm, and the documented trail length is over 2.5 meters. These giant trails are indicative of a well developed nervous system for the coordination of work done by a strong and complex musculature. The mode of feeding and the diet of these animals are not known yet, although some evidence suggests their selective microphagy: the numerous transverse grooves on their ventral side could have been lined with cilia transporting food particles to the medial grove and from there to the slit-like mouth.

New fossil material from the Vendian of the White Sea region led to a new interpretation of Kimberella as a mollusk-like triploblastic animal with a high dorsal, non-mineralized shell. In the course of recent excavations in the valley of the Solza River, Onega Peninsula, short trails have been found behind a few small (1 cm long) specimens of Kimberella. These animals were buried alive by a sand blast in the place of their habitation. While Kimberella was trying to crawl out of the sand cover, its dorsal shell was plunging in an arch-like cross section of the space above the animal, which is preserved in negative hyporelief (concave furrows on the sole of the sandstone bed). The sand layer was saturated with water, and it was thin enough to let these animals continue crawling under conditions of relatively low sedimentation rates. At Winter Coast on the White Sea more common are thick lenses of sandstone corresponding to fast, catastrophic sedimentation which preserved the body imprints of several individuals of Kimberella of various sizes, but no locomotion trace fossils because of instant burial under the heavy load of sediment.

The scratch marks, reported from the Neoproterozoic of South Australia, were originally interpreted as imprints of spicules. These scratch marks as tracks produced by Ediacaran arthropod-like organisms known from rare body fossils, that are, however, still to be described formally. The fine details of the morphology and taphonomy of these fossils from his collection. Recently, these ichnofossils were interpreted as scratches made by the radula of a mollusk. Researchers suggested that the trace fossils should be considered grazing marks of Kimberella.

Scratch marks arranged in a fan-like pattern are a common element of fossil assemblages in the Vendian deposits of the White Sea region. These ichnofossils are preserved in positive relief on the sole of the accompanying bed (sandstone). The stratigraphic and taphonomic cooccurrence of the scratch marks and body fossils of Kimberella is well documented from the Upper Vendian of the White Sea. However, only recently, a few specimens have been collected with the body fossil and scratch marks preserved jointly. These feeding tracks may reflect the repeated functioning of the proboscis carrying two hook-like teeth at its end. Although this bilateral organ might be considered as a predecessor of a true radula, it worked differently: as the animal could stretch and contract the proboscis, it was able to search a vast area of the bottom surface (and 1-2 mm below it) and to rake the food particles (meiofauna and/or algae) and ingest them. Comparison of the body size and the size of the feeding tracks shows that the length of the proboscis could be comparable to that of the body of Kimberella.

The discovery of locomotion and feeding trace fossils preserved together with the body fossils of their producers finally proves that the Ediacara-type organisms are metazoans, which puts the Vendian fossils into their phylogenetic and evolutionary context. The new data represented here leave no space for any non-metazoan interpretations of the Ediacara fossils.

Organic Predecessor of Mineralized Skeletons

The Ediacara fauna of the Vendian is traditionally characterized as consisting of non-skeletal or soft-bodied animals. This is generally true insofar as this concerns heavy mineral skeletons like a mollusk shell or corallite. Most Ediacara-type fossils are preserved as casts and molds of soft-bodied invertebrates without traces of biomineralization. The great majority of the Vendian animals did not possess large mineralized parts, like their recent counterparts do, such as in a few phyla of worms. Nevertheless, there is a growing body of evidence that some forms of Ediacara animals had rather rigid organic structures, both internal and external. These skeletal elements include a dorsal carapace, a shell, teeth-like organs, spicules, and sclerites.

Late Riphean tubular fossils Pararenicola, Proto-arenicola and Parmia as well as the Vendian Saarina, Calyptrina and related forms originally interpreted as annelidomorph metazoans, may well be compared to the theca-like exoskeleton of coelenterate polyps similar to the chitinous periderm tubes of the polyps of some recent coronate (Coronata, Scyphozoa) medusa. One can attribute the slender conical and slightly calcified Cloudina and poppy-like Namacalathus and the recently discovered conulariid Vendoconularia to the same group of external tubular skeletons. Another Ediacara species with a preserved exo-skeleton is Conomedusites, which shows a hard part (theca) supporting a cup-like soft body with marginal tentacles. Long tubular Sabellidites that appears in the uppermost part of the Vendian has a complex wall structure and may represent a higher grade of metazoan organization. The imprints of spicules in an Ediacara sponge.

The bilateral metazoan Andiva described from the siliciclastic deposits of the Vendian deposits of the Winter Coast, White Sea region, Russia, shows bipolarity in its shield-like imprints. Its modes of deformation, growth pattern, and the regeneration marks, though rare, indicate that the animal had a convex dorsal carapace of a triploblastic invertebrate, but the anatomy of which is yet to be discovered. Its thin and flexible carapace was composed of an organic, non-mineralized, and slowly degradeable substance. The fossil Andiva has much in common in its overall body plan, fine structure, and mode of preservation with the well-known Ediacara taxa. Ovatoscutum and Chondroplon which originally interpreted as the bilateral chitinous pneumatophores of the oldest Chondroplidae (Hydrozoa, Coelenterata). According to this interpretation, these two forms have to be considered skeletal remains rather than imprints of soft tissues, similar to the Early Devonian chondrophorine Plectodiscus. However, in the light of new morphological and taphonomic data related to Andiva, the nature of Ovatoscutum, Chondroplon, and Dickinsonia is reconsidered. These taxa form a clade of vagile epibenthic bilaterians with a thin and flexible dorsal carapace.

A rigid dorsal carapace inferred from the high relief of their fossils is also found in Parvancorina, Vendomia, and Onega as well as in some other arthropod-like forms. The elongated and high dorsal shell made of flexible and rigid organic material is documented in the Vendian mollusk-like Kimberella. In the absence of mineralized hard parts, a thin organic skeleton, such as a shell or carapace, is often preserved as an imprint comparable to that of the soft body. Both the carapace and soft body decomposed completely, leaving no primary substance, but it is possible to distinguish one from the another by comparing deformations of various parts of the fossil. In addition, the carapace always keeps its fine sculpture (ribs, suturae) intact even under severe deformation of the body.

The comb-like chitinous Redkinia from the Vendian of the Russian Platform is an example of complex teeth-like structures that resemble scolecodonts or even parts of a sophisticated conodont apparatus. Fan-like sets of scratch marks, associated with and attributed to the grazing activity of Kimberella, have been produced by its proboscis with two sharp teeth at its end. This hook-like tooth made deep, narrow cuts in the muddy sediment or in the bacterial film.

Thus, there is strong evidence for the presence of skeleton and hard parts in the metazoans of the Vendian siliciclastic basins. However, there is no evidence of active biomineralization, or more in particular, of carbonate biomineralization, the most widespread in the Animal Kingdom. This absence of carbonate biomineralization could be explained by analogy to recent cold-water marine basins where carbonate skeletons are rather exceptional. If the organism does have carbonate biominerals, its skeleton is usually extremely thin and is very often protected from the external environment by layers of organic compounds (conchiolin etc.) The shallow-water Antarctic calcareous organisms as a group are always extremely fragile. This group, including rare calcareous foraminifera, prosobranch gastropods, bivalves, scaphopods, and brachiopods, shows dwarfing, that may be reminiscent of the small shelly fossils (SSF) of the Late Vendian and Early Cambrian, though an essential part of those were sclerites of larger invertebrates.

The very slow growth of calcareous organisms and their small size is indicative of the high-energy cost of calcium biomineralization in cold-water environments. The problem of cold-water biochemistry of invertebrates is well beyond the aim of this paper, but it is suggested that calcium biomineralization for the metazoans of the Precambrian cold-water basins was as difficult as for recent ones.

Some calcareous groups may not even have been able to exist in the Vendian, and if they had, those forms should have been of very small size and have had extremely fragile skeletons subjected to quick dissolution in diagenetic and/or postgenetic sedimentary processes. We may therefore never be able to identify any remains of their mineralized skeletons, but we have now firm evidence that skeletonization in the Precambrian invertebrates took place well before the metazoan biomineralization.

From Skeletonization to Biomineralization

An organic skeleton preceded the rise of the mineralized skeleton in most of the existing animal phyla. The oldest known kinds of non-mineralized external skeletons in metazoans are peridermal tubes with very thin, chitin-like walls that might belong to solitary cnidarian-grade polyps (comparable to Vendoconularia). As several authors showed, the present rise in atmospheric CO_2 levels causes significant changes in surface ocean pH and carbonate chemistry that slow down calcification in planktonic organisms, corals and coralline macroalgae. This may explain the absence of calcified eukaryotic organisms through the largest part of the Proterozoic marked by high CO_2 levels in the atmosphere. The low temperatures of the Neoproterozoic glacial periods inhibited biomineralization as well. The Vendian fossil record shows a large variety of organic hard parts in both cold and temperate climatic zones, and a weak biomineralization in carbonate basins.

The almost simultaneous development of phosphate, carbonate and silicate biomineralization in metazoans at the beginning of the Cambrian Period some 545 Ma suggests ecological causes for this phenomenon. Taking into account that the enzymes responsible for the biomineralization are temperature-sensitive, one can suppose that the sudden appearance of invertebrates with a mineralized skeleton was related to global warming or to the colonization of the warm carbonate basins by the early metazoans. Metazoan colonization of the warm carbonate basins and the low-paleolatitude environments might be assisted by the acquisition of endosymbiotic photosynthesizing algae that, in turn, promoted biomineralization through the supply of additional oxygen.

Another cause of the rapid diversification of biomineralized phyla at the beginning of the Cambrian may be related to the growing length of trophic chains, due to the rapidly increasing biodiversity. The concentration of some ions (in particular, Ca, Mg, P and Si) in their cells increased exponentially up the trophic chain. Being byproducts of detoxification, sclerites and spicules, hard mineralized shells and carapaces immediately became important in morphological evolution, in the growing biodiversity, and as well as a factor of intense selection under the growing pressure predators made.

Also taking into the account the low solubility of oxygen in warm water, one can suppose that the metazoans could not colonize the warm carbonate basins until the oxygen level in the atmosphere rose significantly. The lower heat tolerance of metazoans compared with unicellular eukaryotes and bacteria proves to relate to a decrease in oxygen levels in their body fluids, reflecting an excessive oxygen demand at higher temperatures.

Conclusion Impact of Early Metazoans on Global Ecosystem

The early diversification of the Metazoa strongly influenced their evolutionary rates within the clades already present in marine communities. This implies an important role for ecology in fueling the Cambrian explosion extending across many existing kingdoms. Coincident with the Cambrian radiation of marine invertebrates, protistan microfossils doubled in diversity and turnover rates increased by one order of magnitude. The appearance of new kinds of physiology and feeding habits greatly affecting both the biota and the environments accompanied the explosive radiations of invertebrates during the Vendian and Early Cambrian.

The bioturbation of the sediment resulted in its better aeration, which, in turn, allowed its progressive colonization underneath its surface by a broad variety of aerobic organisms. Both this aeration and increasing life activity within the sediment promoted the recycling of metabolites in the marine ecosystems. Yet, bioturbation disrupted the stability of the substrate necessary for the formation of bacterial mats and biogenic structures such as stromatolites.

The rise of biomineralization in the invertebrate phyla resulted in the vertical expansion of benthic animal life above the sediment/water interface, and in the creation (with other non-metazoan groups of eukaryotes) of reefs as a mechanically stable biotope and a very special ecosystem of considerable habitat diversity. The biomineralization of the metazoans formed new kinds of bioclastic deposits and hardgrounds.

The filtration of ocean water by actively filtering organisms had a great impact on the global ocean ecosystem. The feeding habits of recent crustaceans show that during 24 hours one milligram of living weight of the filterers can filter 360 millilitres of water. A volume of water equal to the world's ocean is filtered in half a year, and the largest part of the inhabited ocean (0-500 meter depth) is filtered in only 20 days. These data, together with more recent results, suggest that the rise of active suspension feeding or filteringin metazoans in the Vendian and Cambrian changed the sediment and water habitats radically.

The rise of active filter-feeding organisms such as sponges, some coelenterates, brachiopods, mollusks, arthropods and echinoderms in the Early Cambrian made the ocean water clear and the photic zone deeper. It thus provided additional opportunities for photosynthesizing organisms to occupy lower levels of the water column and deeper benthic environments. The expansion of the photic zone could thus have resulted in better oxygenation of the pelagic and bottom habitats through the activity of chlorophyll-containing organisms. The removal of fine particles from the sea water and their assemblage into pellets must have increased the permeability of the sediment, leading to better aeration and colonization of the subsurface bottom environments and to more rapid oxidation of the buried organic carbon. Pellet transport increased the rate of food supply for the benthic communities.

The increasing length of trophic chains during the Vendian, and more particularly in the Cambrian, decreased the loss of the main biophile elements and of energy from the ecosystems because of more efficient biological recycling. That, in turn, could have led to the global oligotrophication of the ocean waters. This is consistent with the general decrease in buried organic carbon during the Early Cambrian, as well as with the radiation of the Early Cambrian phytoplankton with external processes, spines, ornamentation and a very small cell size. All these morphological peculiarities of the Early Cambrian phytoplankton can be interpreted as a means to develop a very large surface-volume ratio with its advantages in an oligotrophic environment. This agrees with the fact that 70% of the biomass and 80% of the chlorophyll belongs to the picoplankton in the oligotrophic waters of the modern ocean. The oligotrophication of the Early

Cambrian ocean made some of the feeding habits so common among the Vendian metazoans inefficient (for instance, passive trapping of the food particles by the sedentary suspension feeders). This may have caused the elimination of some Ediacara species from the shallow marine habitats in the Early Cambrian or even slightly earlier.

The colonization of the terrestrial environments could well have began with the rise of triploblastic animals capable of active locomotion over or in the sediment. The acquisition of a dorsal carapace promoted this by strengthening the body structure, making muscles work more efficiently and protecting the animals (especially its respiratory organs) from desiccation, and from the deadly ultraviolet light. Evidence of the oldest known grazing under subaerial conditions on tidal flats is the giant trail Climactichnites from the Late Cambrian siliciclastic deposits of North America. The discoveries of large carapace-bearing Vendian bilaterians able to crawl may extend the history of the colonization of the land far back into the late Proterozoic.

CHAPTER–11

Phytolith

INTRODUCTION

A phytolith ("plant stone") is a rigid microscopic body that occurs in many plants. The most common type of phytolith is the silica phytolith, also called opal phytolith. Silica phytoliths vary in size and shape depending on the plant taxon and plant part (stem, leaf, root) in which they (naturally) occur. Grasses, including rice, wild rice, maize, wheat, and other various grains); crop plants such as beans, squashes, gourds, manioc, canna, and arrowroot; palms; as well as numerous tree species are just some of the plants which contain phytoliths. Calcium oxalate phytoliths are another common type, occurring in the stems of cacti and baobabs. Phytoliths are mentioned in the writings of Charles Darwin.

Function

These objects serve a variety of purposes. In many cases, they appear to lend the plant structure and support, much like the spicules in sponges and leather corals. Others serve to make plants distasteful, lending the plant's tissues a grainy or prickly texture. Finally, calcium oxalate phytoliths serve as a reserve of carbon dioxide. Cacti use these as a reserve for photosynthesis during the day when they close their pores to avoid water loss, while baobabs use this property to make their trunks more flame-resistant.

Archaeology

Phytoliths are very robust in nature, and are useful in archaeology, since they can be used to reconstruct the plants present at a site or an area within a site even though the rest of the plant parts have been burned up or dissolved. Because they are made of the inorganic substances silica or calcium oxalate, phytoliths don't decay when the rest of the plant decays over time and can survive in conditions that would destroy organic residues. Phytoliths can provide evidence of both economically important plants and those that are indicative of the environment at a particular time period. They may be extracted from residue on many sources: dental calculus (buildup on teeth); food preparation tools like rocks, grinders, and scrapers; cooking or storage containers; ritual offerings; and garden areas.

Palaeontology

Phytoliths are abundant in the fossil record and have been reported from the late Devonian onwards. They can be used to identify palaeoenvironments and track vegetational change. Occasionally, paleontologists find and identify phytoliths associated with extinct plant-eating animals (e.g., herbivores). Findings such as these reveal useful information about the diet of these extinct animals, and also shed light on the evolutionary history of many different types of plants. Paleontologists in India have recently identified grass phytoliths in dinosaur dung, strongly suggesting that the evolution of grasses began earlier than previously thought.

Japanese and Korean archaeologists refer to grass and crop plant phytoliths as 'plant opal' in archaeological literature.

CARBON SEQUESTRATION

The carbon can be accumulated in many phytoliths, and as such could provide a long term option to sequester carbon in soil in the plant residual silica.

Acritarch

Acritarchs are small organic fossils, present from approximately 2500 million years ago to the present. Their diversity reflects major ecological events such as the appearance of predation and the Cambrian explosion.

Definition

In general, any small, non-acid soluble (i.e. non-carbonate, non-siliceous) organic structure that can not otherwise be accounted for is classified as an acritarch. Acritarchs include the remains of a wide range of quite different kinds of organisms - ranging from the egg cases of small metazoans to resting cysts of many different kinds of chlorophyta (green algae). It is likely that some acritarch species represent the resting stages (cysts) of algae that were ancestral to the dinoflagellates. The nature of the organisms associated with older acritarchs is generally not clear, though many are probably related to unicellular marine algae. In theory, when the biological source (taxon) of an acritarch does become known, that particular microfossil is removed from the acritarchs and classified with its proper group. Whilst the classification of acritarchs into form genera is entirely artificial, it is not without merit, as the form taxa show similar traits to genuine taxa - for example an 'explosion' in the Cambrian and a mass extinction at the end of the Permian.

Occurrence

Acritarchs are found in sedimentary rocks from the present back into the Precambrian. They are typically isolated from siliciclastic sedimentary rocks using hydrofluoric acid but are occasionally extracted from carbonate-rich rocks. They are excellent candidates for index fossils used for dating rock formations in the Paleozoic Era and when other fossils are not available. Because most acritarchs are thought to be marine, they are also useful for palaeoenvironmental interpretation.

Diversity

Acritarchs first appear in rocks about two billion years old, but at about one billion years ago they started to increase in abundance, diversity, size, complexity of shape and especially size and number of spines. Their populations crashed during the Snowball Earth episodes, when all or very nearly all of the Earth's surface was covered by ice or snow, but they proliferated in the Cambrian explosion and reached their highest diversity in the Paleozoic. The increased spininess one billion years ago possibly resulted from the need for defence against predators, especially predators large enough to swallow them or tear them apart. Other groups of small organisms from the Neoproterozoic era also show signs of anti-predator defences.

Further evidence that acritarchs were subject to herbivory around this time comes from a consideration of taxon longevity. The abundance of planktonic organisms that evolved between 1700 and 1400 million years ago was limited by nutrient availability – a situation which limits the origination of new species because the existing organisms are so specialised to their niches, and no other niches are available for occupation. Around about 1,000 million years ago, species longevity fell sharply, suggesting that predation pressure, probably by protist herbivores, became an important factor. Predation would have kept populations in check, meaning that some nutrients were left unused, and new niches were available for new species to occupy.

NEOPROTEROZOIC

The Neoproterozoic Era is the unit of geologic time from 1,000 to 542 +/- 0.3 million years ago. The terminal Era of the formal Proterozoic Eon (or the informal "Precambrian"), it is further subdivided into the Tonian, Cryogenian, and Ediacaran Periods. The most severe glaciation known in the geologic record occurred during the Cryogenian, when ice sheets reached the equator and formed a possible "Snowball Earth"; and the earliest fossils of multicellular life are found in the Ediacaran, including the earliest animals.

At the onset of the Neoproterozoic the supercontinent Rodinia, which had assembled during the late Mesoproterozoic, straddled the equator. During the Tonian, rifting commenced which broke Rodinia into a number of individual land masses. Possibly as a consequence of the low-latitude position of most continents, several large-scale glacial events occurred during the Era including the Sturtian and Marinoan glaciations. These glaciations are believed to have been so severe that there were ice sheets at the equator—a state known as the "Snowball Earth".

The idea of the Neoproterozoic Era came on the scene relatively recently—after about 1960. Nineteenth century paleontologists set the start of multicelled life at the first appearance of hard-shelled animals called trilobites and archeocyathids. This set the beginning of the Cambrian period. In the early 20th century, paleontologists started finding fossils of multicellular animals that predated the Cambrian boundary. A complex fauna was found in South West Africa in the 1920s but was misdated. Another was found in South Australia in the 1940s but was not thoroughly examined until the late 1950s. Other possible early fossils were found in Russia, England, Canada, and elsewhere (see *Ediacaran biota*). Some were determined to be pseudofossils, but others were revealed to be members of rather complex biotas that are still poorly understood. At least 25 regions worldwide yielded metazoan fossils prior to the classical Cambrian boundary.

A few of the early animals appear possibly to be ancestors of modern animals. Most fall into ambiguous groups of frond-like animals(?); discoids that might be holdfasts for stalked animals(?) ("medusoids"); mattress-like forms; small calcaerous tubes; and armored animals of unknown provenance. These were most commonly known as Vendian biota until the formal naming of the Period, and are currently known as Ediacaran biota. Most were soft bodied. The relationships, if any, to modern forms are obscure. Some paleontologists relate many or most of these forms to modern animals. Others acknowledge a few possible or even likely relationships but feel that most of the Ediacaran forms are representatives of (an) unknown animal type(s).

In addition to *Ediacaran biota*, later two other types of biota were discovered in China (the so-called Doushantuo formation and Hainan formation).

TERMINAL PERIOD

The nomenclature for the terminal period of the Neoproterozoic has been unstable. Russian geologists referred to the last period of the Neoproterozoic as the Vendian, and the Chinese called it the Sinian, and most Australians and North

Americans used the name Ediacaran. However, the International Union of Geological Sciences ratified (2004) the Ediacaran age to be a geological age of the Neoproterozoic, ranging from 630 +5/-30 to 542 +/- 0.3 million years ago. The Ediacaran boundaries are the only Precambrian boundaries defined by biologic Global Boundary Stratotype Section and Points, rather than the absolute Global Standard Stratigraphic Ages.

CHAPTER–12

Nannofossils

INTRODUCTION

The numbers of genera and species of microfossils and nannofossils, which have been detected in the Mesozoic sediments along the Dorset coast, are varied and complex. Furthermore classification can be very difficult. The sheer numbers of individual microorganisms on 1 cm^2 of the area of each peel shows that many of these sediments are largely composed of actual biogenic material. These include crinoidal limestone, Kimmeridgian shales, clays and limestone bands, Purbeck micrite and clay beds, Cretaceous chalk and marl layers etc.

Of course individual examination of separated microfossils and nannofossils can be undertaken by crushing, sieving, decanting and treatment with various chemicals, of the original rocks. These procedures are certainly very useful for the classification of many microfossils. However the cellulose lacquer peel technique, discussed here, gives an overall picture of the structure of the various sediments. This applies also to cellulose acetate peels and accurately prepared thin sections.

VARIETY OF NANNOFOSSILS

There is little doubt that the microfossils and nannofossils, present in rocks and sediments, are the largest part of the geological record, in terms of variety and numbers of individual past forms of life. Their study leads to increasing understanding of the Earth's history, past climatic changes, past geographical and geological developments (from plate tectonics and ocean floor spreading), the dating of rock strata, the search for valuable mineral deposits (petroleum exploration) and also a large part of the history of life on our planet.

Furthermore sedimentary rocks contain many microscopic fragments of larger plants and animals, which are worthy of our attention and help to build a general picture of ancient environments (e.g. Mesozoic conifer forests).

Those amongst us who are privileged to observe Nature with an optical microscope, or an electron microscope, can become fully aware of the exquisite complexity of our surroundings. This applies to the study of micropalaeontology, in particular, because the active use of the imagination is required, as well as natural science and accurate observation to build pictures of the remote past ages. Obviously there are many problems and uncertainties to challenge our ideas and theories, and encourage new thinking. This is an age of developing natural science with the aid of rapidly advancing technology and new methods of processing information.

It is hoped that the following illustrations and notes will be of interest to those who wish to observe some of these microscopic objects and features with their own microscopes. The preparation of the cellulose lacquer rock peels does require some practical skill, which can be acquired by "trial and error" methods. To obtain the best results the softer layered rocks should be sampled. The type of peel can be assessed usually with XPL to show the material which has been removed from the rock surface. Likewise XPL can indicate which images on the peel are replicas, and produce no optical figure.

The real diametre of the Coccolithophore is estimated at 11 microns (Fig. 11.1). The surface coccoliths are circa 4.2 microns in diameter. Clearly electron microscope images are needed for closely detailed examination. The species here is believed to be *Ellipsagelosphaera brittanica*. Similar coccolithophore images have been identified on stained cellulose lacquer peels, but these are usually less well defined than on the unstained peels. However, most surprising, some of the stained coccolithophores (and coccoliths) show a deep blue colour due to Ferroan Calcite composition, whereas many others are pale pink in colour suggesting very pure calcite composition. This differential staining of coccolithophores and coccoliths requires confirmation.

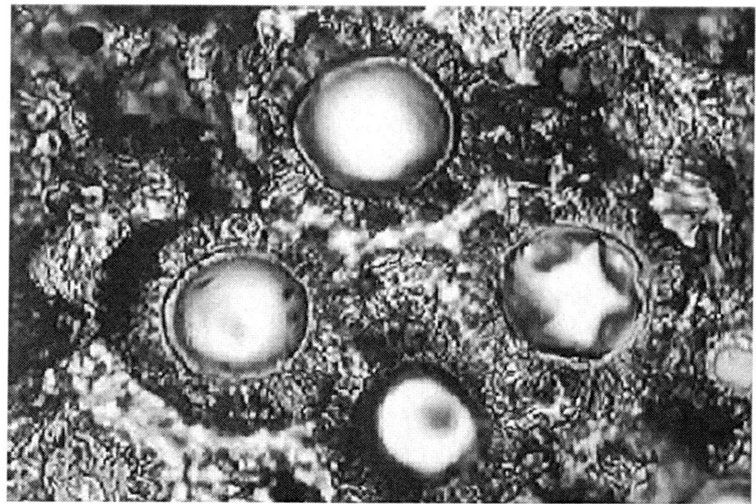

Fig. 11.1: Here we have images of Dinoflagellate 5 cysts from a stained cellulose lacquer peel. The highly magnified picture was derived from the Jurassic, Upper Kimmeridge Clay, Freshwater Steps Stone Band. Map Reference: SY.943.772. Freshwater Steps, west of Hounstout Cliff, Dorset coast. Oil-immersion objective N.A. = 1.25. The total field area of the image = circa 100 microns X 70 microns.

The best images are obtained for classification by the techniques of Palynology, which involves the removal of all the mineral contents of the sediment, using concentrated hydrofluoric acid. This technique is used also for the isolation of pollen grains and plant spores. Slides of the organic material only are prepared in a Canada-balsam mounting, as a strew of microfossils for microscopic examination. An important paper by N.S. Ioannides, G. N. Stravrinos and C. Downie 19768 based on the Kimmeridgian microplankton from Clavell's Hard, Dorset coast, classified some sixty species of Dinoflagellate Cysts and came to some important conclusions about the origins of the kerogen in the Kimmeridge Oil Shale.

Ostracods are microscopic Crustacean arthropods of great geological interest and are frequently studied because of their widespread occurrence in various sediments. As arthropods, living Ostracods are complex multicellular animals with a digestive system, a central nervous system and developed genital organs. The bodies appear unsegmented and are contained in a calcitic carapace consisting of two valves. The microfossils are generally found as carapaces only.

The prominent Ostracod carapace has a maximum diameter of about 813 microns. There are also fragmented carapaces present, as well as a fragment of conifer wood, consisting of removed fusain and replica type imagery. The red colour on the peel is due to stained Micrite. The darker colour may be due to Marl or clay. Lulworth bed 107 in Durleston Bay lies just below the well-known "Cinder Bed", which marks a temporary marine incursion, and is the accepted boundary between the Jurassic and Cretaceous periods. The "Cinder Bed" shows masses of marine Oyster-rich sediment but the beds below it show numerous freshwater fauna, including Ostracods.

This Cypridea Ostracod carapace section has a main axial diameter of about 1210 microns. This lacustrine and brackish water type of Ostrocod is widely distributed in the Purbeck freshwater beds. The Cytheracea Ostracods are found in marine environments. Marl and clay areas are also shown. The peel was examined with brightfield illumination and one polarizing plate above the objective. Some fusain can be seen on this image.

The differential staining of the minute nannofossils on this cellulose lacquer peel are accentuated by the brightfield PPL illumination. This unique flask-shaped object either represents an unknown species (perhaps a type of Calpionellid) or may represent the cell fission of an unkown microorganism.

CHAPTER–13

Stromatolite
(*The Oldest Fossils*)

INTRODUCTION

Stromatolites (from Greek, *strõma*, mattress, bed, stratum, and *lithos*, rock) are layered accretionary structures formed in shallow water by the trapping, binding, and cementation of sedimentary grains by biofilms of microorganisms, especially cyanobacteria (commonly known as blue-green algae).

Morphology

In many respects stromatolites are the most intriguing fossils that are our singular visual portal (except for phylogenetic determination of conserved nucleic acid sequences and molecular fossils) into deep time on earth, the emergence of life, and the eventual evolving of the beautiful life forms from Cambrian to modern time. A small piece of stromatolite encodes biological activity perhaps spanning thousands of years. In broad terms, stromatolites are fossil evidence of the prokaryotic life that remains today, as it has always been, the preponderance of biomass in the biosphere. For those that subscribe to the theory of the living earth, it is the prokaryotes that maintain the homeostasis of the earth, rendering the biosphere habitable for all other life. They maintain and recycle the atomic ingredients upon which proteins that "are" all life are made, including oxygen, nitrogen and carbon. We humans are, in simple terms, bags of water filled with proteins and prokaryotic bacteria (the bacteria in your body outnumber the cells in your body about 10 to 1). We humans have descended from organisms that adapted to living in a prokaryotic world, and we humans retain (conserved in evolutionary terms) in our mitochondria the cellular machinery to power our cells that we inherited (i.e., endosymbiosis) from the prokaryotes of deep time on earth.

Stromatolites and their close cousins the thrombolites, are rock-like buildups of microbial mats that form in limestone- or dolostone-forming environments. Together with oncoids (formerly called "algal biscuits" or "Girvanella"), they typically form by the baffling, trapping, and precipitation of particles by communities of microorganisms such as bacteria and algae. In some cases, they can form inorganically, when seawaters are oversaturated with chemical precipitates. Stromatolites are defined as laminated accretionary structures that have synoptic relief (i.e., they stick up above the seafloor). Stromatolite-building communities include the oldest known fossils, dating back some 3.5 billion years when the environments of Earth were too hostile to support life as we know it today. We can presume that the microbial communities consisted of complex consortia of species with diverse metabolic needs, and that competition for resources and differing motility among them created the intricate structures we observe in these ancient fossils.

Excluding some exceedingly rare Precambrian fossils such as the Ediacaran fauna, stromatolites are the only fossils encoding the first 7/8th of the history of life on earth. They encode the role that ancient microorganisms played in the evolution of life on earth and in shaping earth's environments. The fossil record of stromatolites is astonishingly extensive, spanning 4 billion years of geological history with the forming organisms possibly having occupied every conceivable environment that ever existed. Today, stromatolites are nearly extinct in marine environments, living a precarious existence in only a few localities worldwide. Modern stromatolites were first discovered in Shark Bay, Australia in 1956, and through out western Australia in both marine and non-marine environments. New stromatolite localities have continued to be discovered in various places such as the Bahamas, the Indian Ocean and Yellowstone National Park, to name but a few localities.

Stromatolites are most often described as biogenically-produced structures formed by colonies of photosynthesizing cyanobacteria. However, this is an enormous oversimplification. Science now knows that all domains of life (the Archaeans, Eubacteria, and Eukaryotes) all appeared in the Archaean Era. Which of the prokaryotes came first, the Archaeans or the Eubacteria remains a mystery, but a consensus is emerging that these primitive microorganisms laterally exchanged genes; if so, the concept of the single common ancestor for all life becomes a bit fuzzy. While formation by colonies of cyanobacteria is probably the primary mechanism for formation of stromatolite in the deep time of the Archaean and half way through the Proterozoic, it is unlikely to have been the only mechanism. Recent research indicates the other prokaryotic and the most genetically diverse domain of life, the Archaeans, evolved alongside and possibly swapped genes with the Eubacteria. All prokaryotes (both Eubacteria and the Archaeans reproduce by cell division (binary or multiple fission) and, lacking sex, are essentially clones and among the slowest evolving organisms. Moreover, molecular fossils indicate that primitive Eukaryotic microorganisms appeared more than 3.5 Ba. Thus, before the end of the Archaean time some 2.5 Ba, all three domains of life (Eubacteria, Archaea, and Eukaryotes) existed and were likely already quite diverse. Some were autotrophs, some chemotrophs and some heterotrophs, and collectively they had a multiplicity of metabolic processes from which to derive their energy. Just as microorganisms were extremely diverse in deep time, so is there a corresponding extreme diversity of biogenic and chemical mechanisms that are plausible for the formation of laminar carbonate and other structures that we call stromatolite. Ascribing all stromatolite formation in the Archaean and Proeterozoic to cyanobacteria, seems an unreasonable assumption.

Other research posits, based on genome sequencing, that cyanobacteria may have originated as late as 2.3 billion years ago, and were preceded by sulfur-oxidizing bacteria and sulfate-reducing bacteria. This mirrors the changes in the geochemical record, centered around 2.7 billion years ago. The hypothesis is consistent with geology that finds isotopic fractionation of sulfur compounds becomes large, followed by the sudden increase in oxygen in the atmosphere and surface water environments at about 2.2 or 2.3 Ba.

Whether or not stromatolite contains preserved cellular structure remains controversial, though the consensus is that in special cases, remnants of ancient cell structure can be viewed using special polishing techniques and high magnification. Additionally, determination of atom ratios (so called molecular fossils) in Archaean sediment from Australia has led to sound conjecture that microorganisms with nuclei appeared before 3.8 Ba.

Nonetheless, the cyanobacteria are conjectured to have been the predominant form of life on early earth for more than 2 Billion years, and were likely responsible for the creation of earth's atmospheric oxygen, consuming CO_2 and releasing O_2 through their photosynthetic metabolism. Creation of the modern atmosphere is, of course, perhaps the most critical event in geological history that powered the Cambrian explosion and subsequent evolution of the aerobic forms of life, including all animals. During Precambrian times, bacterial mats formed a platform for trapping and cementation of sediment. For photosynthetic bacteria, depletion of carbon dioxide in the surrounding water could cause precipitation of calcium carbonate that along with grains of sediment were then trapped within the sticky layers of mucilage (that formed a film for UV protection) that surrounding the bacterial colonies. Cyanobacteria are also capable of directly precipitating calcium carbonate, with minimal incorporation of sediment within the structure. The bacteria could repeatedly re-colonize the growing hard sedimentary platform, forming layer upon layer in a cyclic repetitive process. The resulting successive layering can assume a myriad of shapes dependent upon microorganism and environment, and if left undisturbed by forces of nature could form huge domes and flat laminar structures that grew upward toward the life-sustaining rays of the sun.

It is generally conjectured that cyanobacteria were the source of oxidants for banded iron-formation. However, recently resolved phylogenetic trees based on whole genomic DNA sequences show that cyanobacteria were one of the last major lineages to diverge off the bacterial tree. This newly resolved tree shows that sulfur-oxidizing bacteria and sulfate-reducing bacteria arose before cyanobacteria did. This mirrors the changes in the geochemical record, centered around 2.7 billion years ago. At this time, the isotopic fractionation of sulfur compounds becomes large, followed by the sudden increase in oxygen in the atmosphere and surface water environments at about 2.2 or 2.3 billion years ago.

Stromatolites are also variously described as being formed by algae that are, in turn, assumed to be plants; this description still persists in old textbooks and on the Internet, but is scientifically incorrect. It is a holdover from a time that cyanobacteria were thought to be algae (and were called blue-green algae) and from when algae were thought to be plants. Actually, cyanobacteria are prokaryotic bacteria (domain of life Eubacteria), and "genomic" science is sill debating whether eukaryotic, photosynthetic, and autotrophic algae are plants or deserve a distinctive phylogenetic grouping. Regardless, the eukaryotic algae did not appear until about 1.5 Ba, some 2 billion years after stromatolites significantly began forming. It is therefore likely that stromatolite formation by algae was not significant until the Phanerozoic, or possibly the Late Proterozoic.

While not always recognized as such, Banded Iron Formations (BIFs) are another form of stromatolite. BIFs are massive, laterally extensive and globally distributed chemical sediment deposits that consist primarily of Fe-bearing minerals (iron oxides) and silica. Iron can occur naturally in two states. Reduced, or ferric iron is soluble in water. In the Archaean oceans, prodigious ferric iron was released from Earth's interior. In the presence of oxygen, however, the iron becomes oxidized to ferrous iron and precipitates out as a solid. Thus, banded iron layers are the result of oxygen released by photosynthetic organisms combining with dissolved iron in Earth's oceans to form insoluble iron oxides. The banding is assumed to result from cyclic peaks in oxygen production. It is unclear whether these were seasonal or followed some other cycle. It is assumed that initially the Earth started out with vast amounts of iron dissolved in the world's seas. BIFs in the geologic record from 3.8 Ga (Isua, West Greenland) to about 1.8 Ga with a maximal abundance at about 2.5 Ga, and a reoccurrence in Neoproterozoic time (from about 0.8 and 0.6 Ga). The scientific literature commonly attributes the disappearance of BIFs to the fact that deep oceans became oxidized at ~1800 Ma; their formation ostensibly required anoxic deep waters to deliver hydrothermally derived Fe2+ to locations where deposition took place. It provides evidence of an alternative mechanism for BIFs by organisms that directly oxidize Fe(II) as an energy source. It argues that the bacterial genera Gallionella and Chromatium use such metabolism and that both are likely to have existed in Precambrian oceans. Interestingly, it is estimated that the amount of oxygen lock-up in earth's BIFs is some 10 times the amount contained in the atmosphere.

Recent research supports the hypothesis that stromatolite form diversity increased through the Paleoproterozoic, reached a maximum in the Mesaproterozoic at about 1.5 Ba that persisted to about 700 Ma, and steadily declined to several taxa by the Precambrian-Cambrian boundary. This is in contrast to previous data indicating a steep decline at 2 Ba that now appears from the data to be an artifact of 50% of all stromatolite coming from a single basin and author; thus, this decline was in volume. When this regional anomaly is removed, the steepest decline in forms appears to have occurred in the Cambrian. By normalizing stromatolite forms with volume of preserved carbonate rock, the authors posit that the steepest decline in stromatolite form diversity occured in the late Neoproterozoic, and culminated in the Lower Cambrian, coincident with the widespread appearance of macroscopic metazoa and significant bioturbation (i.e., activity of bottom-living animals that keeps sediments oxygenated and homogenous). Conversely, fossil stromatolite indirectly supports the hypothesis that the diversification of major animal phyla occurred between 1 and 1.2 Ba. Since laminated sediments are a sign of oxygen depletion in the bottom zone of the sea, bioturbation would inhibit the building of stromatolitic structures. This research seems consistent with evolutionary theory that would anticipate diversification of stromatolite forming taxa due to selective pressure from other organisms that were emerging and themselves diversifying.

We will likely have no more than a sketchy understanding of the paleoenvironments in which stromatolite was formed in the deep time Precambrian, and only an incomplete understanding of the environments in the Paleozoic. Sound conjecture is possible if we examine the now rare environments that support stromatolitic growth during modern times. Cyanobacteria are found to be a primary organism in the formation of microbial carbonates. These prokaryotic bacteria (slang name is blue-green algae owning to pigmentation involved in photosynthesis) are now only found in areas where there is reduced grazing and burrowing by other organisms, and a low occurrence of macro-algae and plants. Environments where modern stromatolites are found typically are hypersaline, but also include areas of high alkalinity, low nutrients, high or low temperatures, and strong wave or current actions. The obvious pattern emerges that modern stromatolites tend to exist in areas that most other life forms consider less desirable or possibly intolerable. Thus, organisms producing modern stromatolite are generally limited to areas where organisms with which they have to compete and/or organisms that might use them for nutrients are not prevalent.

A variety of stromatolite morphologies exist including conical, stratiform, branching, domal and columnar types. Stromatolites occur widely in the fossil record of the Precambrian, but are rare today. Very few ancient stromatolites contain fossilized microbes. While features of some stromatolites are suggestive of biological activity, others possess features that are more consistent with abiotic (non-organic) precipitation. Finding reliable ways to distinguish between biologically-formed and abiotic (non-biological) stromatolites is an active area of research in geology.

Fossil Record

Stromatolites were much more abundant on the planet in Precambrian times. While older, Archean fossil remains are presumed to be colonies of single-celled blue-green bacteria, younger (that is, Proterozoic) fossils may be primordial forms of the eukaryote chlorophytes (that is, green algae). One genus of stromatolite very common in the geologic record is *Collenia*. The earliest stromatolite of confirmed microbial origin dates to 2,724 million years ago.

Stromatolites are a major constituent of the fossil record for about the first 3.5 billion years of life on earth, with their abundance peaking about 1,250 million years ago. They subsequently declined in abundance and diversity, which by the start of the Cambrian had fallen to 20% of their peak. The most widely-supported explanation is that stromatolite builders fell victims to grazing creatures: implying that sufficiently complex organisms were common over one billion years ago.

The connection between grazer and stromatolite abundance is well documented in the younger Ordovician evolutionary radiation; stromatolite abundance also increased after the end-Ordovician and end-Permian extinctions decimated marine animals, falling back to earlier levels as marine animals recovered.

Stromatolites at Lake Thetis, Western Australia

While prokaryotic cyanobacteria themselves reproduce asexually through cell division, they were instrumental in priming the environment for the evolutionary development of more complex eukaryotic organisms. Cyanobacteria are thought to be largely responsible for increasing the amount of oxygen in the primeval earth's atmosphere through their continuing photosynthesis.

Cyanobacteria use water, carbon dioxide, and sunlight to create their food. The byproducts of this process are oxygen and calcium carbonate (lime). A layer of mucus often forms over mats of cyanobacterial cells. In modern microbial mats, debris from the surrounding habitat can become trapped within the mucus, which can be cemented together by the calcium carbonate to grow thin laminations of limestone. These laminations can accrete over time, resulting in the banded pattern common to stromatolites. The domal morphology of biological stromatolites is the result of the vertical growth necessary for the continued infiltration of sunlight to the organisms for photosynthesis.

MODERN OCCURRENCE

Modern stromatolites are mostly found in hypersaline lakes and marine lagoons where extreme conditions due to high saline levels exclude animal grazing. One such location is Hamelin Pool Marine Nature Reserve, Shark Bay in Western Australia where excellent specimens are observed today, and another is Lagoa Salgada, state of Rio de Janeiro, Brazil, where modern stromatolites can be observed as bioherm (domal type) and beds. Inland stromatolites can also be found in saline waters in Cuatro Ciénegas, a unique ecosystem in the Mexican desert. Modern stromatolites are only known to prosper in an open marine environment in the Exuma Cays in the Bahamas.

Layered *spherical* growth structures named oncolites are similar to stromatolites, and are also known from the fossil record.

CHAPTER–14

Cyanobacteria

INTRODUCTION

Domain : Bacteria

Phylum : Cyanobacteria

Cyanobacteria, also known as blue-green algae, blue-green bacteria or Cyanophyta, is a phylum of bacteria that obtain their energy through photosynthesis. The name "cyanobacteria" comes from the color of the bacteria (Greek: (kyanós) = blue). They are a significant component of the marine nitrogen cycle and an important primary producer in many areas of the ocean, but are also found on land.

Stromatolites of fossilized oxygen-producing cyanobacteria have been found from 2.8 billion years ago The ability of cyanobacteria to perform oxygenic photosynthesis is thought to have converted the early reducing atmosphere into an oxidizing one, which dramatically changed the composition of life forms on Earth by provoking an explosion of biodiversity and leading to the near-extinction of oxygen-intolerant organisms. Chloroplasts in plants and eukaryotic algae have evolved from cyanobacteria via endosymbiosis.

FORMS

Cyanobacteria are found in almost every conceivable environment, from oceans to fresh water to bare rock to soil.

Most are found in fresh water, while others are marine, occur in damp soil, or even temporarily moistened rocks in deserts. A few are endosymbionts in lichens, plants, various protists, or sponges and provide energy for the host. Some live in the fur of sloths, providing a form of camouflage.

Cyanobacteria include unicellular and colonial species. Colonies may form filaments, sheets or even hollow balls. Some filamentous colonies show the ability to differentiate into several different cell types: vegetative cells, the normal, photosynthetic cells that are formed under favorable growing conditions; akinetes, the climate-resistant spores that may form when environmental conditions become harsh; and thick-walled heterocysts, which contain the enzyme nitrogenase, vital for nitrogen fixation. Heterocysts may also form under the appropriate environmental conditions (anoxic) wherever nitrogen is necessary. Heterocyst-forming species are specialized for nitrogen fixation and are able to fix nitrogen gas, which cannot be used by plants, into ammonia (NH_3), nitrites (NO_2^-) or nitrates (NO_3^-), which can be absorbed by plants and converted to protein and nucleic acids. The rice paddies of Asia, which produce about 75% of the world's rice, could not do so were it not for healthy populations of nitrogen-fixing cyanobacteria in the rice paddy fertilizer too.

Many cyanobacteria also form motile filaments, called hormogonia, that travel away from the main biomass to bud and form new colonies elsewhere. The cells in a hormogonium are often thinner than in the vegetative state, and the cells on either end of the motile chain may be tapered. In order to break away from the parent colony, a hormogonium often must tear apart a weaker cell in a filament, called a necridium.

Each individual cell of a cyanobacterium typically has a thick, gelatinous cell wall. They differ from other gram-negative bacteria in that the quorum sensing molecules autoinducer-2 and acyl-homoserine lactones are absent. They lack flagella, but hormogonia and some unicellular species may move about by gliding along surfaces. In water columns some cyanobacteria float by forming gas vesicles, like in archaea.

PHOTOSYNTHESIS

Cyanobacteria have an elaborate and highly organized system of internal membranes which function in photosynthesis. Photosynthesis in cyanobacteria generally uses water as an electron donor and produces oxygen as a by-product, though some may also use hydrogen sulfide as occurs among other photosynthetic bacteria. Carbon dioxide is reduced to form carbohydrates via the Calvin cycle. In most forms the photosynthetic machinery is embedded into folds of the cell membrane, called thylakoids. The large amounts of oxygen in the atmosphere are considered to have been first created by the activities of ancient cyanobacteria. Due to their ability to fix nitrogen in aerobic conditions they are often found as symbionts with a number of other groups of organisms such as fungi (lichens), corals, pteridophytes (Azolla), angiosperms (*Gunnera*) etc.

Cyanobacteria are the only group of organisms that are able to reduce nitrogen and carbon in aerobic conditions, a fact that may be responsible for their evolutionary and ecological success. The water-oxidizing photosynthesis is accomplished by coupling the activity of photosystem (PS) II and I (Z-scheme). In anaerobic conditions, they are also able to use only PS I — cyclic photophosphorylation — with electron donors other than water (hydrogen sulfide, thiosulphate, or even molecular hydrogen) just like purple photosynthetic bacteria. Furthermore, they share an archaeal property, the ability to reduce elemental sulfur by anaerobic respiration in the dark. Their photosynthetic electron transport shares the same compartment as the components of respiratory electron transport. Actually, their plasma membrane contains only components of the respiratory chain, while the thylakoid membrane hosts both respiratory and photosynthetic electron transport.

Attached to thylakoid membrane, phycobilisomes act as light harvesting antennae for the photosystems. The phycobilisome components (phycobiliproteins) are responsible for the blue-green pigmentation of most cyanobacteria. The variations to this theme is mainly due to carotenoids and phycoerythrins which give the cells the red-brownish coloration. In some cyanobacteria, the color of light influences the composition of phycobilisomes. In green light, the cells accumulate more phycoerythrin, whereas in red light they produce more phycocyanin. Thus the bacteria appear green in red light and red in green light. This process is known as complementary chromatic adaptation and is a way for the cells to maximize the use of available light for photosynthesis.

A few genera, however, lack phycobilisomes and have chlorophyll *b* instead (*Prochloron*, *Prochlorococcus*, *Prochlorothrix*). These were originally grouped together as the prochlorophytes or chloroxybacteria, but appear to have developed in several different lines of cyanobacteria. For this reason they are now considered as part of cyanobacterial group.

RELATIONSHIP TO CHLOROPLASTS

Chloroplasts found in eukaryotes (algae and plants) likely evolved from an endosymbiotic relation with cyanobacteria. This endosymbiotic theory is supported by various structural and genetic similarities. Primary chloroplasts are found among the green plants, where they contain chlorophyll *b*, and among the red algae and glaucophytes, where they contain phycobilins. It now appears that these chloroplasts probably had a single origin, in an ancestor of the clade called Primoplantae. Other algae likely took their chloroplasts from these forms by secondary endosymbiosis or ingestion.

It was once thought that the mitochondria in eukaryotes also developed from an endosymbiotic relationship with cyanobacteria; however, it is now suspected that this evolutionary event occurred when aerobic bacteria were engulfed by anaerobic host cells. Mitochondria are believed to have originated not from cyanobacteria but from an ancestor of *Rickettsia*.

CYANOBACTERIA AND EARTH HISTORY

The biochemical capacity to use water as the source for electrons in photosynthesis evolved once, in a common ancestor of extant cyanobacteria. The geologic record indicates that this transforming event took place early in our planet's history, at least 2450-2320 million years ago (Ma), and possibly much earlier. Geobiological interpretation of Archean (>2500 Ma) sedimentary rocks remains a challenge; available evidence indicates that life existed 3500 Ma, but the question of when oxygenic photosynthesis evolved continues to engender debate and research. A clear paleontological window on cyanobacterial evolution opened about 2000 Ma, revealing an already diverse biota of blue-greens. Cyanobacteria remained principal primary producers throughout the Proterozoic Eon (2500-543 Ma), in part because the redox structure of the oceans favored photautotrophs capable of nitrogen fixation. Green algae joined blue-greens as major primary producers on continental shelves near the end of the Proterozoic, but only with the Mesozoic (251-65 Ma) radiations of dinoflagellates, coccolithophorids, and diatoms did primary production in marine shelf waters take modern form. The most common cyanobacterial structures in the fossil record include stromatolites and oncolites. Cyanobacteria remain critical to marine ecosystems as primary producers in oceanic gyres, as agents of biological nitrogen fixation, and, in modified form, as the plastids of marine algae.

CYANOBACTERIAL EVOLUTION FROM COMPARATIVE GENOMICS

Recent high-throughput sequencing has provided DNA sequences at an unprecedented rate, posing considerable analytical challenges, but also offering insight into the genetic mechanisms of adaptation. Here we present a comparative genomics-based approach towards understanding the evolution of these mechanisms in cyanobacteria. Historically, systematic methods of defining morphological traits in cyanobacteria have posed a major barrier in reconstructing their true evolutionary history. The advent of protein, then DNA, sequencing - most notably the use of 16S ribosomal RNA as a molecular marker - helped circumvent this barrier and now forms the basis of our understanding of the history of life on Earth. However, these tools have proved insufficient for resolving relationships between closely related cyanobacterial species. The 24 cyanobacteria whose genomes have been compared occupy a wide variety of environmental niches and play major roles in global carbon and nitrogen cycles. By integrating phylogenetic data inferred for hundreds to nearly 1000 protein coding genes common to all or most cyanobacteria, we are able to reconstruct an evolutionary history of the entire phylum, establishing a framework for resolving how their metabolic and phenotypic diversity came about.

CLASSIFICATION

The cyanobacteria were traditionally classified by morphology into five sections, referred to by the numerals I-V. The first three - Chroococcales, Pleurocapsales, and Oscillatoriales - are not supported by phylogenetic studies. However, the latter two - Nostocales and Stigonematales - are monophyletic, and make up the heterocystous cyanobacteria. The members of Chroococales are unicellular and usually aggregated in colonies. The classic taxonomic criterion has been the cell morphology and the plane of cell division. In Pleurocapsales, the cells have the ability to form internal spores (baeocytes). The rest of the sections include filamentous species. In Oscillatorialles, the cells are uniseriately arranged and do not form specialized cells (akinets and heterocysts). In Nostocalles and Stigonematalles the cells have the ability to develop heterocysts in certain conditions. Stigonematales, unlike Nostocalles include species with truly branched trichome. Most taxa included in the phylum or division Cyanobacteria have not yet been validly published under the Bacteriological Code. Except:

- The classes Chroobacteria Hormogoneae and Gloeobacteria
- The orders Chroococcales, Gloeobacterales, Nostocales, Oscillatoriales, Pleurocapsales and Stigonematales
- The families Prochloraceae and Prochlorotrichaceae
- The genera Halospirulina, Planktothricoides, Prochlorococcus, Prochloron, Prochlorothrix.

BIOTECHNOLOGY AND APPLICATIONS

Certain cyanobacteria produce cyanotoxins like anatoxin-a, anatoxin-as, aplysiatoxin, cylindrospermopsin, domoic acid, microcystin LR, nodularin R (from *Nodularia*), or saxitoxin. Sometimes a mass-reproduction of cyanobacteria results in algal blooms.

The unicellular cyanobacterium *Synechocystis* sp. PCC6803 was the third prokaryote and first photosynthetic organism whose genome was completely sequenced. It continues to be an important model organism The smallest genomes have been found in *Prochlorococcus* spp. (1.7 Mb) and the largest in *Nostoc punctiforme* (9 Mb). Those of *Calothrix* spp. are estimated at 12-15 Mb, as large as yeast.

Some cyanobacteria are sold as food, notably *Aphanizomenon flos-aquae* and *Arthrospira platensis* (Spirulina). It has been suggested that they could be a much more substantial part of human food supplies, as a kind of superfood.

Along with algae, some hydrogen producing cyanobacteria are being considered as an alternative energy source, notably at Oregon State University, in research supported by the U.S. Department of Energy, Princeton University, Colorado School of Mines, Ohio University as well as at Uppsala University, Sweden.

HEALTH RISKS

Some species of cyanobacteria produce neurotoxins, hepatotoxins, cytotoxins, and endotoxins, making them dangerous to animals and humans. Several cases of human poisoning have been documented but a lack of knowledge prevents an accurate assessment of the risks.

ARCHEAN

The Archean also spelled Archaean, formerly called the Archaeozoic, also spelled (Archeozoic or Archæozoic) is a geologic eon before the Proterozoic and Paleoproterozoic, before 2.5 Ga (billion years ago, or 2,500 Ma). Instead of being based on stratigraphy, this date is defined chronometrically. The lower boundary (starting point) has not been officially recognized by the International Commission on Stratigraphy, but it is usually set to 3.8 Ga, at the end of the Hadean eon. In older literature, the Hadean is included as part of the Archean.The name comes from the ancient Greek (Arkhe), meaning "beginning, origin".

Archean Earth

At the beginning of the Archean, the Earth's heat flow was nearly three times higher than it is today, and was still twice the current level by the beginning of the Proterozoic (2,500 Ma). The extra heat may have been remnant heat from the planetary accretion, partly heat of formation of the iron core, and partially caused by greater radiogenic heat production from short-lived radionuclides such as uranium-235.

The majority of Archean rocks which still survive are metamorphic and igneous rocks. Volcanic activity was considerably higher than today, with numerous hot spots, rift valleys, and eruption of lavas including unusual types such as komatiite. Nevertheless, intrusive igneous rocks predominate throughout the crystalline cratonic remnants of the Archean crust which survive today. These are magmas which infiltrated into host rocks, but solidified before they could erupt at the Earth's surface. Examples include great melt sheets and voluminous plutonic masses of granite, diorite, ultramafic to mafic layered intrusions, anorthosites and monzonites known as sanukitoids.

The Earth of the early Archean may have had a different tectonic style. Some scientists think that because the Earth was hotter, plate tectonic activity was more vigorous than it is today, resulting in a much greater rate of recycling of crustal material. This may have prevented cratonisation and continent formation until the mantle cooled and convection slowed down. Others argue that the subcontinental lithospheric mantle was too buoyant to subduct, and that the lack of Archean rocks is a function of erosion by subsequent tectonic events. The question of whether or not plate tectonic activity existed in the Archean is an active area of modern geoscientific research.

There were no large continents until late in the Archean: small protocontinents were the norm, prevented from coalescing into larger units by the high rate of geologic activity. These felsic protocontinents probably formed at hotspots rather than subduction zones, from a variety of sources: igneous differentiation of mafic rocks to produce intermediate and felsic rocks, mafic magma melting more felsic rocks and forcing granitization of intermediate rocks, partial melting of mafic rock, and from the metamorphic alteration of felsic sedimentary rocks. Such continental fragments may not have been preserved unless they were buoyant enough or fortunate enough to avoid energetic subduction zones.

An explanation for the general lack of Hadean rocks (older than 3800 Ma) is the amount of extrasolar debris present within the early solar system. Even after planetary formation, considerable volumes of large asteroids and meteorites still existed, and bombarded the early Earth until approximately 3800 Ma. A barrage of particularly large impactors known as the late heavy bombardment may have prevented any large crustal fragments from forming by literally shattering the early protocontinents.

Archean Palaeoenvironment

The Archean atmosphere apparently lacked free oxygen. Temperatures appear to have been near modern levels even within 500 Ma of Earth's formation, with liquid water present, as evidenced by certain highly deformed gneisses produced by metamorphism of sedimentary protoliths. Astronomers think that the sun was about one-third dimmer than at present, which may have contributed to lower global temperatures than otherwise expected. This is thought to reflect larger amounts of greenhouse gases than later in the Earth's history.

By the end of the Archaean c. 2600 Mya, plate tectonic activity may have been similar to that of the modern Earth. There are well-preserved sedimentary basins, and evidence of volcanic arcs, intracontinental rifts, continent-continent collisions and widespread globe-spanning orogenic events suggesting the assembly and destruction of one and perhaps several supercontinents. Liquid water was prevalent, and deep oceanic basins are known to have existed by the presence of banded iron formations, chert beds, chemical sediments and pillow basalts.

Archean Geology

Although a few mineral grains are known that are Hadean, the oldest rock formations exposed on the surface of the Earth are Archean or slightly older. Archean rocks are known from Greenland, the Canadian Shield, the Baltic shield, Scotland, India, Brazil, western Australia, and southern Africa. Although the first continents formed during this eon, rock of this age makes up only 7% of the world's current cratons; even allowing for erosion and destruction of past formations, evidence suggests that continental crust equivalent to only 5-40% of the present amount formed during the Archean

In contrast to the Proterozoic, Archean rocks are often heavily metamorphized deep-water sediments, such as graywackes, mudstones, volcanic sediments, and banded iron formations. Carbonate rocks are rare, indicating that the oceans were more acidic due to dissolved carbon dioxide than during the Proterozoic. Greenstone belts are typical Archean formations, consisting of alternating units of metamorphosed mafic igneous and sedimentary rocks. The meta-igneous rocks were derived from volcanic island arcs, while the metasediments represent deep-sea sediments eroded from the neighboring island arcs and deposited in a forearc basin. Greenstone belts represent sutures between protocontinents

Archean Life

Fossils of cyanobacterial mats (stromatolites) are found throughout the Archean, becoming especially common late in the eon, while a few probable bacterial fossils are known from chert beds. In addition to the domain Bacteria (once known as Eubacteria), microfossils of the domain Archaea have also been identified.

Life was probably present throughout the Archean, but may have been limited to simple non-nucleated single-celled organisms, called Prokaryota (formerly known as Monera). There are no known eukaryotic fossils, though they might have evolved during the Archean without leaving any fossils. No fossil evidence yet exists for ultramicroscopic intracellular replicators such as viruses.

Cyanobiont

A cyanobiont is a cyanobacterium that lives in symbiosis with an eukaryote, sometimes inside the cells of the eukaryote.

The cyanobiont fixes nitrogen, and sometimes also performs photosynthesis for the host organism. The cyanobiont is often vertically transmitted, this is the case for cyanobionts of diatoms, lichens, didemnid ascidians, *Azolla* ferns and some sponges. Some of these cyanobionts cannot live without the host organism.

Hypolith

In Arctic and Antarctic ecology, a hypolith is a photosynthetic organism that lives underneath rocks in climatically extreme deserts such as Cornwallis Island and Devon Island in the Canadian high Arctic. The community itself is the hypolithon.

Hypolithons are protected from harsh ultraviolet radiation and wind scouring by their rock, which can also trap moisture. The rocks are generally translucent to allow for the penetration of light. Writing in *Nature,* ecologist Charles S. Cockell of the British Antarctic Survey and Dale Stokes describe how hypoliths reported to date (until 2004) had been found under quartz, which is one of the most common translucent rocks.

However, Cockell reported that on Cornwallis Island and Devon Island, 94-95% of a random sample of 850 opaque dolomitic rocks were colonized by hypoliths, and found that the communities were dominated by cyanobacteria. The rocks chosen were visually indistinguishable from those nearby, and were about 10cm across; the hypolithon was visible as a greenish coloured band. Cockell proposed that rock sorting by periglacial action, including that during freeze-thaw cycles, improves light penetration around the edges of rocks (see granular material and Brazil nut effect).

Cockell and Stokes went on to estimate the productivity of the Arctic communities by monitoring the uptake of sodium bicarbonate labelled with Carbon-14 and found that (for Devon Island) productivity of the hypolithon was comparable to that of plants, lichens, and bryophytes combined (0.8 ± 0.3 g m^{-2} y^{-1} and 1 ± 0.4 g m^{-2} y^{-1} respectively) and concluded that the polar hypolithon may double previous estimates of the productivity of that region of the rocky polar desert.

Oxygen Catastrophe

The Oxygen Catastrophe was a massive environmental change believed to have happened during the Siderian period at the beginning of the Paleoproterozoic era of the Precambrian, about 2.4 billion years ago. It is also called the Oxygen Crisis, Oxygen Revolution, or The Great Oxidation.

When evolving lifeforms developed oxyphotosynthesis about 3.5 billion years ago, molecular oxygen was initially produced in limited quantities. With time, this oxygen accumulated and eventually caused an ecological crisis to the biodiversity of the time, as oxygen was toxic to the microscopic anaerobic organisms dominant then.

However, this transforming change also provided a new opportunity for biological diversification, as well as tremendous changes in the nature of chemical interactions between rocks, sand, clay, and other geological substrates and the Earth's air, oceans, and other surface waters. Despite natural recycling of organic matter, life had remained energetically limited until the widespread availability of oxygen. This breakthrough in metabolic evolution greatly increased the free energy supply to living organisms, having a truly global environmental impact.

Proterozoic

The Proterozoic is a geological eon representing a period before the first abundant complex life on Earth. The name Proterozoic comes from the Greek "earlier life." The Proterozoic Eon extended from 2500 Ma to 542.0 ± 1.0 Ma (million years ago), and is the most recent part of the old, informally named 'Precambrian' time.

The Proterozoic consists of 3 geologic eras, from oldest to youngest:

- Paleoproterozoic;
- Mesoproterozoic; and
- Neoproterozoic.

The well-identified events were:

- The transition to an oxygenated atmosphere during the Mesoproterozoic.
- Several glaciations, including the hypothesized Snowball Earth during the Cryogenian period in the late Neoproterozoic.
- The Ediacaran Period (635 to 542 Ma) which is characterized by the evolution of abundant soft-bodied multicellular organisms.

Index

F

G

H

I

R